GAIN-OF-FUNCTION RESEARCH

Summary of the Second Symposium
March 10–11, 2016

Piers Millett, Jo Husbands, Frances Sharples, and Audrey Thevenon,
Rapporteurs

Board on Life Sciences
Division on Earth and Life Studies

Board on Health Sciences Policy
Health and Medicine Division

Committee on Science, Technology, and Law
Policy and Global Affairs Division

The National Academies of
SCIENCES · ENGINEERING · MEDICINE

THE NATIONAL ACADEMIES PRESS
Washington, DC
www.nap.edu

THE NATIONAL ACADEMIES PRESS 500 Fifth Street, NW Washington, DC 20001

This activity was supported by Contract No. HHSN263201200074 between the National Academy of Sciences and the National Institutes of Health, by the John D. and Catherine T. MacArthur Foundation, and by the Wellcome Trust. Any opinions, findings, conclusions, or recommendations expressed in this publication do not necessarily reflect the views of any organization or agency that provided support for the project.

International Standard Book Number-13: 978-0-309-44077-6
International Standard Book Number-10: 0-309-44077-7
Digital Object Identifier: 10.17226/23484

Additional copies of this workshop summary are available for sale from the National Academies Press, 500 Fifth Street, NW, Keck 360, Washington, DC 20001; (800) 624-6242 or (202) 334-3313; http://www.nap.edu.

Printed in the United States of America

Suggested citation: National Academies of Sciences, Engineering, and Medicine. 2016. *Gain-of-Function Research: Summary of the Second Symposium, March 10-11, 2016.* Washington, DC: The National Academies Press. doi: 10.17226/23484.

The National Academies of
SCIENCES · ENGINEERING · MEDICINE

The **National Academy of Sciences** was established in 1863 by an Act of Congress, signed by President Lincoln, as a private, nongovernmental institution to advise the nation on issues related to science and technology. Members are elected by their peers for outstanding contributions to research. Dr. Ralph J. Cicerone is president.

The **National Academy of Engineering** was established in 1964 under the charter of the National Academy of Sciences to bring the practices of engineering to advising the nation. Members are elected by their peers for extraordinary contributions to engineering. Dr. C. D. Mote, Jr., is president.

The **National Academy of Medicine** (formerly the Institute of Medicine) was established in 1970 under the charter of the National Academy of Sciences to advise the nation on medical and health issues. Members are elected by their peers for distinguished contributions to medicine and health. Dr. Victor J. Dzau is president.

The three Academies work together as the **National Academies of Sciences, Engineering, and Medicine** to provide independent, objective analysis and advice to the nation and conduct other activities to solve complex problems and inform public policy decisions. The Academies also encourage education and research, recognize outstanding contributions to knowledge, and increase public understanding in matters of science, engineering, and medicine.

Learn more about the National Academies of Sciences, Engineering, and Medicine at **www.national-academies.org.**

PLANNING COMMITTEE ON GAIN-OF-FUNCTION RESEARCH: THE SECOND SYMPOSIUM

Members

HARVEY V. FINEBERG (*Chair*), President, Gordon and Betty Moore Foundation, San Francisco, California
RONALD M. ATLAS, Professor, University of Louisville, Kentucky
RUTH L. BERKELMAN, Rollins Professor and Director, Center for Public Health Preparedness and Research, Emory University, Atlanta, Georgia
BARRY R. BLOOM, Harvard University Distinguished Service Professor and Joan L. and Julius H. Jacobson Professor of Public Health, Harvard University, Cambridge, Massachusetts
DONALD S. BURKE, Dean, Graduate School of Public Health, and Director, Center for Vaccine Research, University of Pittsburgh, Pennsylvania
PHILIP DORMITZER, Vice President and Chief Scientific Officer: Viral Vaccines, Pfizer Vaccine Research and Development, Pearl River, New York
BARUCH FISCHHOFF, Howard Heinz University Professor, Carnegie Mellon University, Pittsburgh, Pennsylvania
CHARLES N. HAAS, Betz Professor of Environmental Engineering, Drexel University, Philadelphia, Pennsylvania
MICHELLE M. MELLO, Professor of Law, Stanford Law School, and Professor of Health Research and Policy, Stanford University School of Medicine, Stanford, California
SIR JOHN SKEHEL, Emeritus Research Fellow, National Institute for Medical Research, London, United Kingdom

Staff

FRANCES E. SHARPLES, Director, Board on Life Sciences
JO L. HUSBANDS, Scholar/Senior Project Director, Board on Life Sciences
AUDREY THEVENON, Associate Program Officer, Board on Life Sciences
AANIKA SENN, Senior Project Assistant, Board on Life Sciences
JENNA OGILVIE, Research Associate, Board on Life Sciences
BRENDAN McGOVERN, Senior Project Assistant, Water Science and Technology Board

Acknowledgments

This workshop summary has been reviewed in draft form by individuals chosen for their diverse perspectives and technical expertise. The purpose of this independent review is to provide candid and critical comments that will assist the institution in making its published workshop summary as sound as possible and to ensure that the summary meets institutional standards for objectivity, evidence, and responsiveness to the study charge. The review comments and draft manuscript remain confidential to protect the integrity of the process. We wish to thank the following individuals for their review of this summary:

Thomas W. Armstrong, TWA8HR Occupational Hygiene Consulting, LLC
Harvey Fineberg, Gordon and Betty Moore Foundation
Matthew Frieman, University of Maryland
Diane E. Griffin, Johns Hopkins Bloomberg School of Public Health
Charles Haas, Drexel University
Gregory Koblentz, George Mason University
Peter Palese, Icahn School of Medicine at Mount Sinai
David Relman, Stanford University
Robert Webster, St. Jude Children's Research Hospital

Although the reviewers listed above have provided many constructive comments and suggestions, they did not see the final draft of the workshop summary before public release. The review of this summary was overseen by Ellen Wright Clayton, Vanderbilt University, and Michael J. Imperiale, University of Michigan. They were responsible for making certain that an independent examination of this summary was carried out in accordance with institutional procedures and that all review comments were carefully considered. Responsibility for the final content of this summary rests entirely with the rapporteurs and the institution.

Contents

Acronyms and Abbreviations

APLU	Association of Public and Land-grant Universities
ASPR	Office of the Assistant Secretary for Preparedness and Response
BMBL	*Biosafety in Microbiological and Biomedical Laboratories* (CDC and NIH manual)
BWC	Biological Weapons Convention
DURC	dual use research of concern
EASAC	European Academies Science Advisory Council
EHS	environmental, health, and safety
EU	European Union
FACA	Federal Advisory Committee Act
FDA	Food and Drug Administration
GAO	Government Accountability Office
GEC	German Ethics Council
GOF	gain of function/gain-of-function
GOFRC	GOF research of concern
HHS	Department of Health and Human Services

IBC	Institutional Biosafety Committee
IDA	Institute for Defense Analyses
IRB	Institutional Review Board
IRE	Institutional Review Entity
MENA	Middle East and North Africa
MERS-CoV	Middle East respiratory syndrome coronavirus
NAS	National Academy of Sciences
NBACC	National Biodefense Analysis and Countermeasures Center
NIAID	National Institute of Allergy and Infectious Diseases
NIH	National Institutes of Health
NRC	National Research Council
NSABB	National Science Advisory Board for Biosecurity
NSABB WG	National Science Advisory Board for Biosecurity Working Group
OCS	Office of the Chief Scientist, Food and Drug Administration
OSP	Office of Science Policy, National Institutes of Health
OSTP	Office of Science and Technology Policy
PAPR	powered air purifying respirator
PI	principal investigator
R-BAT	Risk–Benefit Assessment Team
RAC	Recombinant DNA Advisory Committee
RBA	risk–benefit assessment
SARS-CoV	severe acute respiratory syndrome coronavirus
UPMC	University of Pittsburgh Medical Center
WHO	World Health Organization

1

Introduction

On March 10-11, 2016, the National Academies of Sciences, Engineering, and Medicine held a public symposium on potential U.S. government policies for the oversight of gain-of-function (GOF) research.[1] This was the Academies' second meeting held at the request of the U.S. government to provide a mechanism to engage the life sciences community and the broader public and solicit feedback on optimal approaches to ensure effective federal oversight of GOF research as part of a broader U.S. government deliberative process. Approximately 125 people attended the event in person, while more than 200 others watched the webcast.[2]

The first symposium, held in December 2014, examined the underlying scientific and technical questions surrounding the potential risks and

[1] As described in the Draft Working Paper of the National Science Advisory Board for Biosecurity, "the phrase 'gain-of-function research' has become synonymous with certain studies that enhance the ability of pathogens to cause disease. However, gain-of-function studies, as well as loss-of-function studies, are common in molecular and microbiology and form the foundation of microbial genetics. Changes to the genome of an organism, whether naturally occurring or directed through experimental manipulations in the laboratory, can manifest as altered phenotypes as biological functions are lost or gained. Such loss- and gain-of-function experiments allow investigators to understand the complex nature of host-pathogen interactions that underlie transmission, infection, and pathogenesis and can help attribute biological function to genes and proteins. The term 'gain-of-function' is generally used to refer to changes resulting in the enhancement or acquisition of new biological functions or phenotypes" (NSABB, 2015a: 7).

[2] The archived webcast, the presentation slides, and a complete transcript of the symposium are available on the project website at http://dels.nas.edu/Upcoming-Event/Gain-Function-Research-Second/AUTO-9-61-70-Q?bname=bls.

benefits of GOF research involving pathogens with pandemic potential (NRC, 2015).[3]

The second symposium focused on discussion of the draft recommendations regarding GOF research of the National Science Advisory Board for Biosecurity's (NSABB's) Working Group (WG). The recommendations are contained in a draft paper, which was released in December 2015 and discussed at an NSABB meeting on January 7-8, 2016 (NSABB, 2015a).[4] It also included discussion of a formal risk and benefit assessment (RBA) commissioned to inform the NSABB's work (Gryphon Scientific, 2015) and sessions devoted to current U.S. policy and the international developments that provide essential context for U.S. decisions. The public symposium did not attempt to develop consensus recommendations, but rather sought individual perspectives and robust discussion to inform the development of the NSABB's final recommendations. The Statement of Task for the symposium may be found in Box 1-1.

This report has been prepared by the rapporteurs as a factual summary of what occurred during the symposium. The planning committee's role was limited to organizing and convening the workshop. The views contained in the report are those of individual workshop participants and do not necessarily represent the views of all workshop participants, the planning committee, or the Academies. The report offers a summary of the key issues and ideas identified during the symposium, but it offers no consensus conclusions or recommendations and is intended to reflect the discussions during the meeting. In order to be as responsive to the charge as possible, it is organized thematically rather than chronologically so that ideas raised at various points in the symposium are grouped together. A complete transcript that provides additional information about the contents of the presentations and discussions is available on the project website.[5]

OPENING REMARKS

The symposium was opened by Ralph J. Cicerone, president of the National Academy of Sciences (NAS). His remarks reflected on the long

[3] In addition to the summary report of the meeting, the archived webcast and the presentation slides may be found at http://dels.nas.edu/Workshop-Summary/Potential-Risks-Benefits-Gain/21666?bname=bls.

[4] The NSABB WG's paper, along with the commissioned papers, the archived webcast, and all the presentations at the January meeting, are available on the NSABB website at http://osp.od.nih.gov/office-biotechnology-activities/event/2016-01-07-130000-2016-01-08-220000/national-science-advisory-board-biosecurity-nsabb-meeting.

[5] The transcript may be found at http://dels.nas.edu/Upcoming-Event/Gain-Function-Research-Second/AUTO-9-61-70-Q?bname=bls.

BOX 1-1
Statement of Task

An ad hoc committee established by the National Research Council (NRC)[a] will organize two public symposia. The first symposium was held on December 15-16, 2014, and included discussion of the following topics:

- Principles important for, and key considerations in, the design of risk and benefit assessments of gain-of-function (GOF) research.
- Potential benefits of the research, including generating new scientific knowledge about viruses with pandemic potential, informing public health responses to a potential pandemic, supporting surveillance efforts to identify possible pandemic strains and provide more time for preparedness, and facilitating the development of vaccines and antiviral therapeutics.
- Potential risks associated with the research, in particular those related to biosafety and biosecurity.
- Alternative methods that may be employed to yield similar scientific insights and/or potential benefits, while reducing potential risks.

The second symposium—the focus of this task—to be held in early 2016, will focus on discussions of the National Science Advisory Board for Biosecurity (NSABB) draft recommendations regarding GOF research. This meeting will also include discussions of the results of the commissioned risk–benefit assessment as well as risk interpretation and analysis to inform decision making. This symposium will provide a mechanism to both engage the life sciences community as well as solicit feedback on optimal approaches to ensure effective federal oversight of GOF research. Of note, the public symposium should not include the development of consensus recommendations, but rather should elicit individual perspectives and robust discussion on the topics described above. Discussions at this symposium will inform the development of the NSABB's final recommendations.

The committee appointed by the NRC to organize and plan the second symposium will develop the symposium agenda, select and invite speakers and discussants, and moderate the discussions. Invited attendees should have a diverse range of perspectives and expertise, including but not limited to public health, biosafety, public health surveillance, research, risk assessment experts, public policy makers, security, and drug and vaccine development; the agenda should also include experts from regions of the world where pathogens with pandemic potential are endemic and from regions of the world conducting GOF research on such pathogens. This 2-day symposium will be webcast and the presentations and background materials will be archived online.

---

[a] On July 1, 2015, the institutional designation became the National Academies of Sciences, Engineering, and Medicine.
SOURCE: NIH, 2015.

history of the NAS's engagement with the complexities of balancing the risks and benefits of science and technology. Providing a neutral forum in which to discuss the scientific underpinnings of complex and controversial topics is one of the major missions of the Academies, and he urged participants to engage fully in the discussions over the two days of the symposium.

Margaret Hamburg, Foreign Secretary of the National Academy of Medicine, then discussed the evolution of oversight of so-called dual use research in the life sciences, from the 2004 report on *Biotechnology Research in an Age of Terrorism* to the current GOF discussions (NRC, 2004).[6] Dr. Hamburg highlighted the role of the Academies in providing science advice to government. She indicated the importance of the GOF debate and the international nature of the issues and diseases involved. Dr. Hamburg noted that while the discussions at the symposium were focused on advice for the U.S. government, they would have implications for the global research enterprise. She underscored the importance of the symposium and its role in building on a wide range of earlier discussions on policy frameworks and approaches to addressing GOF research. This meeting, according to Dr. Hamburg, was an opportunity to look at those frameworks and approaches and identify desirable next steps. She identified a need to develop a strategic approach to support scientific progress while addressing the impacts for our societies.

Jo Handelsman from the White House Office of Science and Technology Policy believed this to be a landmark meeting, one that could direct future policy in important ways. She noted that the White House has focused on issues around GOF research for 18 months and recognizes the need to keep life sciences vibrant but also to protect safety and security across the globe. Officials had become engaged because of concerns around the creation of new pathogens, especially those with pandemic potential. The White House has also worked to address safety incidents at laboratories that raised public concerns over work with such pathogens (Holdren and Monaco, 2014). In response to these concerns, in October 2014 the White House announced a deliberative process and, along with it, a pause on federal funding for certain types of GOF research (White House, 2014a). Dr. Handelsman highlighted the importance of key exceptions to the funding pause to enable necessary emergency research to continue.

The NSABB was asked to draft recommendations for a conceptual approach for dealing with GOF research that would then be made available for public comment. As mentioned above, the Academies were asked

[6] In this context, "dual use" refers to the dilemma that "the same technologies can be used legitimately for human betterment and misused for bioterrorism" (NRC, 2004: 1).

to convene two public meetings to facilitate a broad discussion of all the relevant issues: one to review technical developments, and a second to discuss the draft recommendations prepared by the NSABB as well as policy options for GOF. Dr Handelsman noted that the NSABB's draft recommendations would be revised in light of the discussions at this symposium and in line with the public input they have received. Following this, an interagency process led by the Office of Science and Technology Policy will produce a policy that will provide federal oversight for GOF research and replace the funding pause.

Carrie Wolinetz from the National Institutes of Health began her remarks by stating that a robust life sciences research endeavor is critical to promoting public health and well-being in light of evolving threats posed by microbial pathogens. This endeavor will entail a certain amount of risk, she noted, requiring a thoughtful approach to reducing risk while taking advantage of the broad range of benefits. She commented that GOF research was a fundamental scientific tool to:

- Help define the nature of host–pathogen interactions;
- Enable assessment of the pandemic potential of emerging infectious agents;
- Inform public health and preparedness efforts; and
- Further medical countermeasure development.

Dr. Wolinetz stated that some GOF experiments had raised safety and security concerns about whether they could result in engineered pathogens capable of causing a pandemic if accidentally or deliberately released. There was also concern that information describing their development could be used by those with malign intent to cause harm through a deliberate release.

Dr. Wolinetz described the GOF deliberative process (see Figure 1-1). She recalled that the deliberative process included a pause in funding for GOF research involving influenza viruses and Middle East respiratory syndrome coronavirus (MERS-CoV) and severe acute respiratory syndrome coronavirus (SARS-CoV).[7] She highlighted the role that had been

[7] "New USG [U.S. government] funding will not be released for gain-of-function research projects that may be reasonably anticipated to confer attributes to influenza, MERS, or SARS viruses such that the virus would have enhanced pathogenicity and/or transmissibility in mammals via the respiratory route. The research funding pause would not apply to characterization or testing of naturally occurring influenza, MERS, and SARS viruses, unless the tests are reasonably anticipated to increase transmissibility and/or pathogenicity. In parallel, we will encourage the currently-funded USG and non-USG funded research community to join in adopting a voluntary pause on research that meets the stated definition" (White House, 2014a: 2).

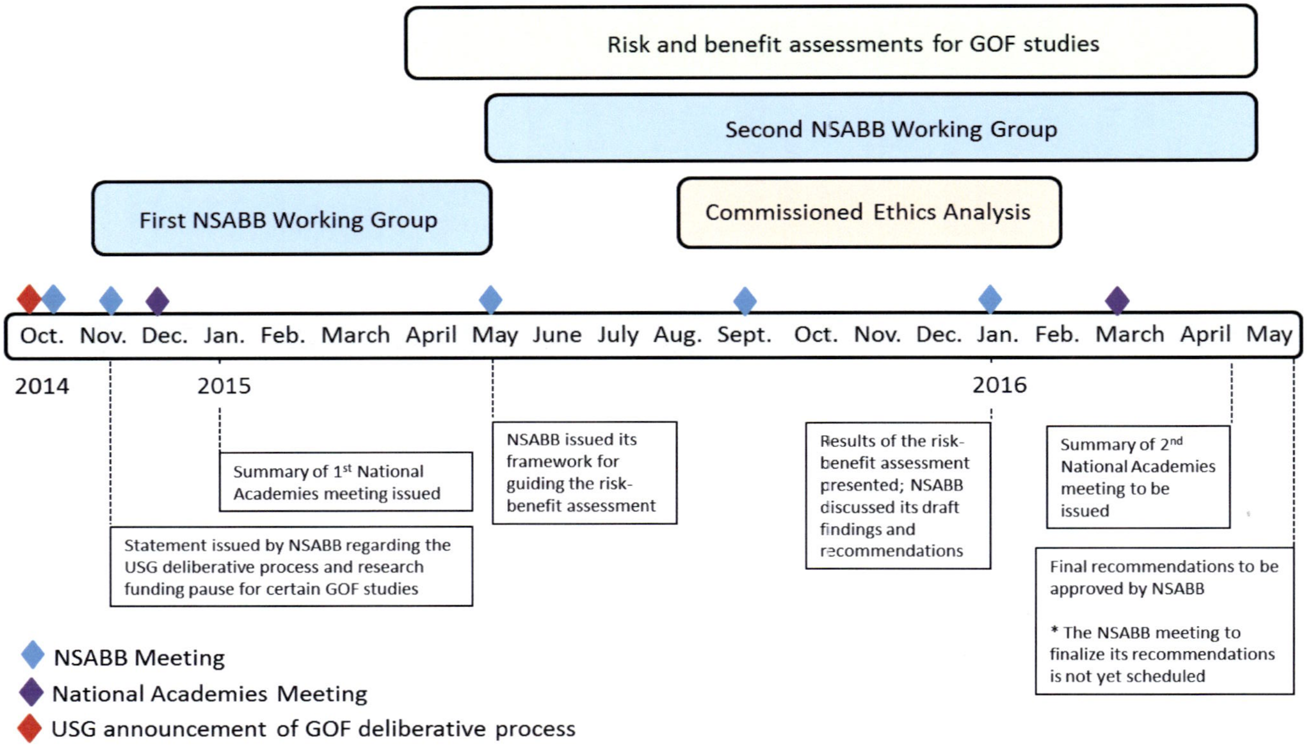

FIGURE 1-1 Timeline of major events in the gain-of-function deliberative process.
NOTE: GOF = gain of function; NSABB = National Science Advisory Board for Biosecurity; USG = U.S. government.
SOURCE: NSABB 2015a: 9.

played by the NSABB and recalled that it had been charged to advise on the design, development, and conduct of a risk–benefit assessment of GOF studies as well as to provide formal recommendations to the U.S. government on the conceptual approach to the evaluation of proposed GOF studies. During the process, the NSABB had also acted as a convening body. Dr. Wolinetz noted that the NSABB had received many valuable inputs to assist it in its work, including the report from the first Academies GOF symposium (NRC, 2015), the risk and benefit assessment conducted by Gryphon Scientific (Gryphon Scientific, 2015), and the ethics report commissioned from Michael Selgelid (Selgelid, 2015). Dr. Wolinetz concluded by noting that more input was being sought, for example, through the discussions at this symposium.

2

The Draft National Science Advisory Board for Biosecurity Policy Framework, the Risk and Benefit Assessment, and Insights for the Policy Process

OVERVIEW OF THE NATIONAL SCIENCE ADVISORY BOARD FOR BIOSECURITY DRAFT WORKING PAPER

Samuel Stanley, the chair of the National Science Advisory Board for Biosecurity (NSABB), highlighted the valuable role played by the National Academies of Sciences, Engineering, and Medicine's first symposium on gain-of-function (GOF) research in the deliberations of the NSABB's Working Group on GOF Issues (hereafter, NSABB WG), in particular during the development of its Draft Working Paper and recommendations. Dr. Stanley reviewed the activities undertaken by the NSABB since the start of the deliberative process. He reviewed the charge to the NSABB and highlighted the outputs produced to date, including *Framework for Conducting Risk and Benefit Assessments of Gain-of-Function Research* in May 2015 (NSABB, 2015b) and *Working Paper Prepared by the NSABB Working Group on Evaluating the Risks and Benefits of Gain-of-Function Studies to Formulate Policy Recommendations* in December 2015 (NSABB, 2015a).

Dr. Stanley introduced the NSABB WG's Draft Working Paper, noting that it included guiding principles for NSABB deliberations; analysis and interpretation of the formal risk and benefit assessment (RBA); consideration of ethical values and decision-making frameworks; analysis of the current policy landscape and potential policy options; preliminary findings from the NSABB WG's analyses; draft recommendations for the NSABB's consideration; and a number of important questions for further consideration. He reviewed key inputs into the work of the NSABB WG.

Dr. Stanley provided some reflections on the RBA prepared by Gryphon Scientific (Gryphon Scientific, 2015), describing it as rigorous and comprehensive and representing a monumental amount of work. The scope of the RBA addressed biosafety risks and biosecurity risks as well as benefits from GOF research. The study had allowed the NSABB to understand the different risks associated with research involving relevant pathogens and certain GOF experiments. It had helped them to identify and distinguish GOF studies that raise significant concerns from those that do not. Dr. Stanley indicated it assisted in identifying and evaluating the potential benefits of GOF studies and in comparing the potential benefits derived from GOF studies to those that may be achieved through alternative approaches.

Drawing on the ethics report prepared by Michael Selgelid (Selgelid, 2015), Dr. Stanley highlighted a number of important values to consider when evaluating research proposals involving GOF studies as well as when establishing mechanisms to review and/or make funding decisions about them. These included both substantive values (such as non-malfeasance, beneficence, social justice, respect for persons, scientific freedom, and responsible stewardship) and procedural values (such as public participation and democratic deliberation, accountability, and transparency).

He noted that there are multiple policies and frameworks already in place for managing risks during the research lifecycle (see Figure 2-1). These include reviews of the scientific merit of proposed research; measures for biosafety oversight, such as the *Biosafety in Microbiological and Biomedical Laboratories* manual (CDC and NIH, 2007) and the *NIH Guidelines for Research Involving Recombinant or Synthetic Nucleic Acid Molecules* (NIH, 2013); the Federal Select Agent Program; the U.S. government policies for federal and institutional oversight of life sciences dual use research of concern (White House, 2012, 2014b); the Department of Health and Human Services framework for guiding funding decisions about certain GOF studies with highly pathogenic avian influenza (HHS, 2012); and measures that relate to sharing and communicating scientific findings and research products. Dr. Stanley noted that the success of these measures depends on effective compliance and implementation. He noted that there were different levels of oversight depending on what pathogen was involved and what was being done with it.

Dr. Stanley then summarized the key findings and recommendations from the NSABB WG's Draft Working Paper. The five key findings are:

1. There are many types of GOF studies and not all of them have the same level of risks. Only a small subset of GOF studies—GOF studies of concern—entail risks that are potentially significant enough to warrant additional oversight;
2. The U.S. government has effective policy frameworks in place for managing risks associated with life sciences research (see Figure 2-1).

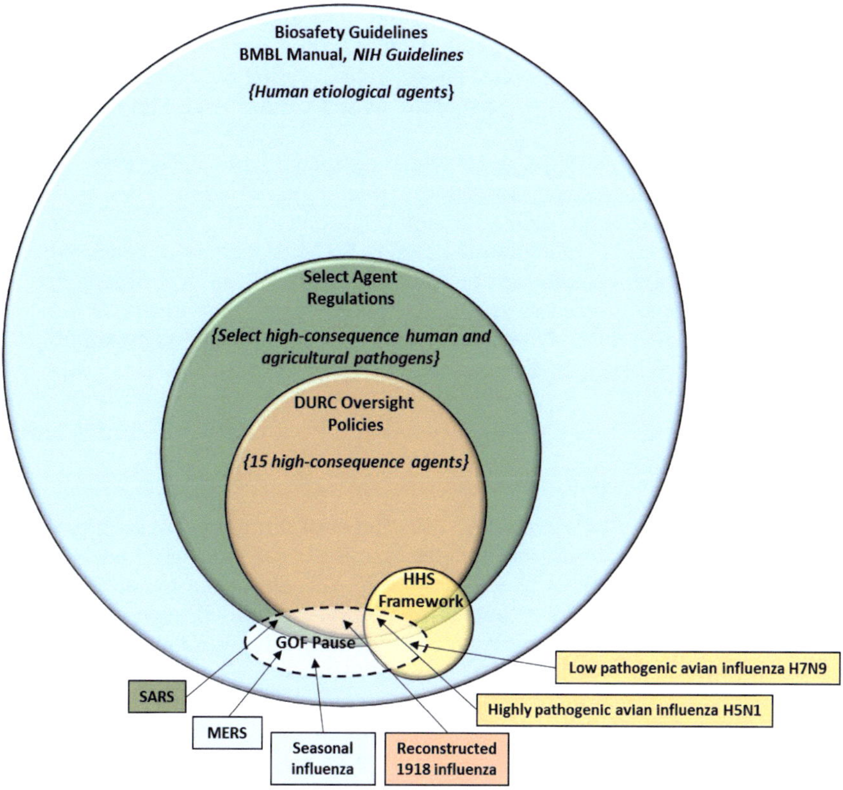

FIGURE 2-1 U.S. government policy frameworks for managing risks associated with life sciences research.
NOTE: BMBL = *Biosafety in Microbiological and Biomedical Laboratories*; DURC = dual use research of concern; GOF = gain-of-function; HHS = Department of Health and Human Services; MERS = Middle East respiratory syndrome; NIH = National Institutes of Health; SARS = severe acute respiratory syndrome.
SOURCE: National Science Advisory Board for Biosecurity, 2015a: 27.

There are several points throughout the research life cycle where, if the policies are implemented effectively, risks can be managed and oversight of GOF studies could be applied;

3. Oversight policies vary in scope and applicability, therefore, current oversight is not sufficient for all GOF studies that raise concern;
4. There are life sciences research studies that should not be conducted on ethical or public health grounds if the potential risks associated with the study are not justified by the potential benefits. Decisions about whether GOF research of concern should be permitted will

entail an assessment of the potential risks and anticipated benefits associated with the individual experiment in question. The scientific merit of a study is a central consideration during the review of proposed studies but other considerations and values are also important; and

5. The biosafety and biosecurity issues associated with GOF studies are similar to those issues associated with all high containment research, but a small subset of GOF studies have the potential to generate strains with high and potentially unknown risks. Managing risks associated with all high containment research requires Federal-level oversight, institutional awareness and compliance, and a commitment by all stakeholders to safety and security. Biosafety and biosecurity are international issues requiring global engagement. (NSABB, 2015a: 3-4)

The NSABB WG's Draft Working Paper also includes four recommendations:

1. Research proposals involving GOF studies of concern entail the greatest risks and should be reviewed carefully for biosafety and biosecurity implications, as well as potential benefits, prior to determining whether they are acceptable for funding. If funded, such projects should be subject to ongoing oversight at the federal and institutional levels;
2. In general, oversight mechanisms for GOF studies of concern should be incorporated into existing policy frameworks. The risks associated with some GOF research of concern can be identified and adequately managed by existing policy frameworks if those policies are implemented properly. However, the level of oversight provided by existing frameworks varies by pathogen. For some pathogens, existing oversight frameworks are robust and additional oversight mechanisms should generally not be required. For other pathogens, existing oversight frameworks are less robust and may require supplementation. All relevant policies should be implemented appropriately and enhanced when necessary to effectively manage risks;
3. The risk-benefit profile for GOF studies of concern may change over time and should be re-evaluated periodically to ensure that the risks associated with such research is adequately managed and the benefits are being realized.
4. The U.S. government should continue efforts to strengthen biosafety and biosecurity, which will foster a culture of responsibility that will support not only the safe conduct of GOF research of concern but of all research involving pathogens. (NSABB, 2015a: 4-5)

A key issue related to the first finding and recommendation is the question of what constitutes "GOF studies of concern." As Dr. Stanley explained:

> GOF research of concern would be a study that can be anticipated to generate a pathogen that is, one, highly transmissible in a relevant mammalian model, two, highly virulent in a relevant mammalian model, and three, is likely more capable of being spread among human populations than currently circulating strains of the pathogen. The first two characteristics are intended to involve the concept of the threshold. That is, the generated pathogen would need to be highly transmissible and highly virulent. Studies of pathogens with moderate virulence and transmissibility entail risks of course, but in general, those risks can be managed through existing mechanisms. The third criterion is intended to capture the concept of pandemic potential. That is, a pathogen could spread rapidly among human populations, either because there's no population immunity, no available counter-measures, or for some other reason. (Stanley, 2016)

The question of the appropriate criteria for defining GOF studies of concern was a recurring theme in subsequent discussions.

Dr. Stanley went on to explain that the NSABB WG had also identified a number of principles for guiding funding decisions related to GOF studies of concern (see Box 2-1).

To further assist in determining how such arrangements might function in practice, the NSABB WG had continued to develop the conceptual approach for the review, funding, and oversight of GOF studies of concern, including a new diagram (see Figure 2-2), which Dr. Stanley presented at the symposium. It includes activities to be undertaken at the institutional and federal levels and details what additional steps would be required for GOF studies of concern. He added that, as discussed at the NSABB's January meeting and in the NSABB WG's Draft Working Paper, the NSABB had highlighted a number of questions that needed further consideration and input (NSABB, 2015a: 46). He said that the NSABB WG was also considering a new question: "What type of body should be tasked with the high-level review of GOF research of concern. Would a FACA-like[1] committee be desirable, or as now envisioned by NSABB, can such reviews be accomplished by federal agencies, or other groups internal to the United States government?"

Dr. Stanley concluded by saying that the NSABB would continue working on its recommendations, with plans for a meeting scheduled for May 24, at which the final report would be discussed and possibly voted

[1] "The Federal Advisory Committee Act [FACA] was enacted in 1972 to ensure that advice by the various advisory committees formed over the years is objective and accessible to the public. The Act formalized a process for establishing, operating, overseeing, and terminating these advisory bodies" (General Services Administration, http://www.gsa.gov/portal/category/21242).

BOX 2-1
NSABB Principles to Guide Funding Decisions for Gain-of-Function Research of Concern

The following principles should guide the review of and funding decisions about research proposals anticipated to involve GOF research of concern:

i. The research proposal has been evaluated by a peer-review process and determined to be scientifically meritorious and has been assessed to be likely to exert a sustained, powerful influence on the research field(s) involved.
ii. An assessment of the overall potential risks and benefits associated with the project determines that the potential risks compared to the potential benefits are justified.
iii. There are no feasible, equally efficacious alternative methods to address the same scientific question in a manner that poses less risk than does the proposed approach.
iv. The investigator and institution proposing the research have the demonstrated capacity to carry it out safely and securely.
v. The research information is anticipated to be broadly and legally shared in order to realize its potential benefits to global health.
vi. The research will be supported through funding mechanisms that include appropriate oversight of (a) all aspects of the research including its conduct, (b) the sharing of data and materials, and (c) the communication of the research.
vii. The proposed research is ethically justifiable.

SOURCE: NSABB, 2015a: 43.

on. He encouraged the participants to continue to submit comments to the NSABB and to take an active role in the symposium discussions.

Harvey Fineberg, chair of the Symposium Planning Committee, then moderated an open discussion of the NSABB WG's Draft Working Paper. In his introductory remarks, Dr. Fineberg highlighted the importance of determining what does, or should, qualify as GOF studies of concern and the subset of research that may warrant additional oversight. He recalled the three characteristics identified by the NSABB WG and stressed the importance of defining the threshold between research of concern and other studies. He commented that the proposed definition did not take into account the starting point of virulence, transmissibility, or resistance of the pathogen. "If you have a very resistant organism that is very virulent if contracted, and all you want to do is to test whether the function of transmissibility could be enhanced, why would that be less of concern than starting with a less virulent, less resistant, less transmissible organism, and trying to produce increased function along all three

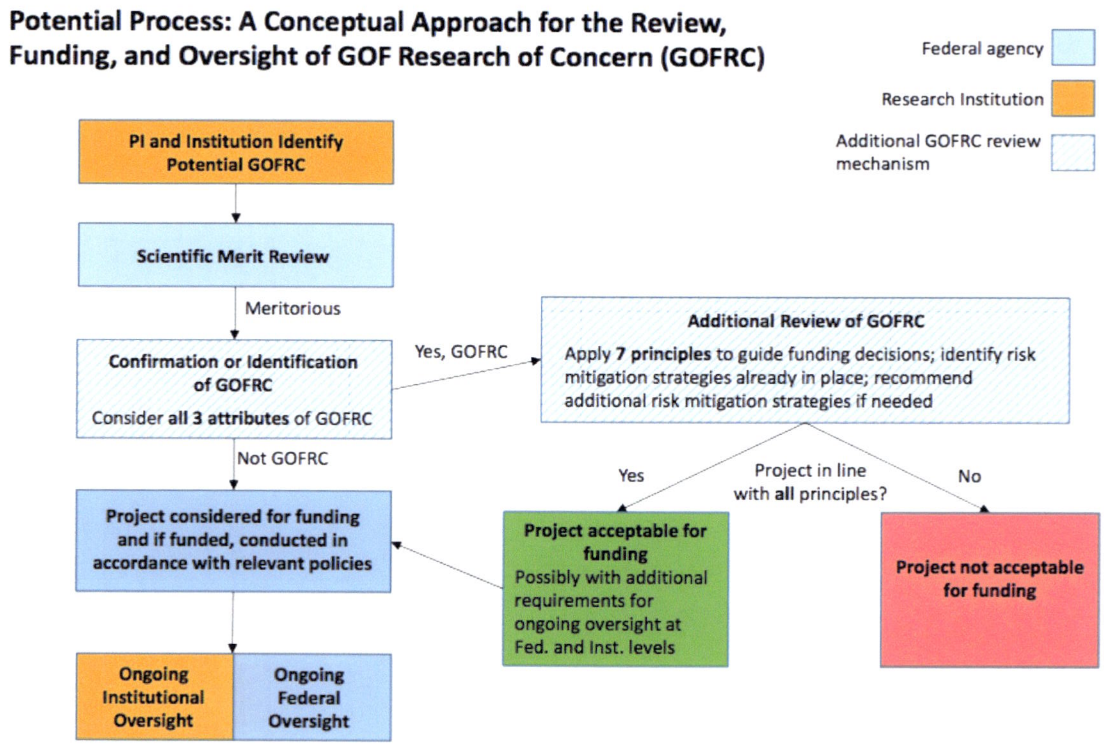

FIGURE 2-2 A conceptual approach for the review, funding, and oversight of gain-of-function research of concern.
NOTE: GOF = gain-of-function; PI = principal investigator.
SOURCE: Stanley, 2016.

dimensions?" He suggested that conceptually it might make more sense to think about "zones" of GOF research where concerns arise, because any combination of the three, or any one, two, or three, leads to a zone of concern outside of what the native organism represents (Fineberg, 2016).

Dr. Fineberg commented that the research enterprise is generally positive because it reveals truths of nature. But there could be a class of investigations that provoke scientific, ethical, or social concerns. In such cases, he felt, the burden of proof as to the value of a specific piece of research would move to those wanting to pursue it.

In conclusion, he added to Margaret Hamburg's comments, noting that, although the NSABB was focused on recommendations for U.S. policy, this was intrinsically a global challenge. He expressed the hope that during the symposium participants would consider the issues related to the development of a global regime to manage this class of research of concern, in addition to and beyond any national regime.

Discussion

Joseph Kanabrocki from the University of Chicago and Kenneth Berns from the University of Florida, co-chairs of the NSABB WG, joined Dr. Stanley on stage for the discussion.

The discussion that followed highlighted several themes from the presentations. The scope of assessing risks and benefits was explored, with questions raised by George Gao from the Chinese Academy of Sciences and the Chinese Center for Disease Control and Prevention about why there was comparatively little focus on loss-of-function experiments, given the difficulty of predicting which method would reduce, or increase, or enhance a virus's virulence or transmissibility. More broadly, Keiji Fukuda from the World Health Organization noted connections to other issues in which life sciences research lies at the heart of safety or security concerns, such as food security or genetically modified organisms. He highlighted the importance of engaging with access and benefit-sharing regimes.

Alternative approaches to tying increased oversight to funding were discussed. Some participants felt that direct regulatory approaches would be preferable. A review of different policy options from a recently published commentary was presented by Thomas Inglesby from the Center for Health Security, University of Pittsburgh Medical Center (UPMC), that included lifting the moratorium on GOF research; seeking an international consensus; securing national and international agreement to restrict the performance of GOF studies of concern; designating a board; establishing clear red lines for GOF studies of concern; and requiring the purchase by research institutions of specific liability insurance policies (Lipsitch et al., 2016). Megan Palmer from Stanford University reflected that several of the questions identified by the NSABB WG as requiring further consideration corresponded to some of the tasks given to the NSABB by the White House. She also asked the NSABB WG to provide key lessons on the limitations of expertise or limitations in the process that might be fed into broader or future discussions on the oversight of life science research.

The importance of international collaboration was stressed and the potential for those wishing to undertake GOF studies of concern to relocate to less restrictive environments was noted by Abdulaziz Alagaili from King Saud University in Saudi Arabia. He also argued that any oversight frameworks should apply to the private sector as well as academia. Piers Millett from Biosecure suggested that an international component should be a significant aspect of future work, including the allocation of necessary resources; a mandate for long-term, sustained engagement; and genuine two-way conversations (rather than the presentation of a finalized solution). He also suggested that any international discussions should be co-hosted by relevant health and security entities to prevent perceptions of bias.

The outcomes from discussions held in other countries about GOF research were highlighted. Filippa Lentzos from King's College London, for example, in a comment made via the Web, noted findings from these discussions of the

- Lack of clear and convincing justifications for GOF studies of concern;
- Role of personal or institutional interests in agenda setting;
- Global dimension of GOF research of concern and the need for an international solution;
- Potential for accidents, abuse, and malpractice, and the intricate relationship between trust and accountability;
- Instability of political contexts and changing security environments, and the need for transparency in biodefense-related research; and
- Need for clear red lines on the most dangerous GOF experiments that apply to the public, private, and military sectors.

She also raised the issue of how to ensure that the lay public's voice is heard and incorporated into the decision-making process around GOF of concern research.

Another participant, Catherine Rhodes from the University of Cambridge, recalled a recent meeting in the United Kingdom in which influenza researchers indicated an interest in developing international approaches to the oversight of relevant research but feared any such process becoming dominated by existing U.S. policy discussions.

On the question of how to define "GOF studies of concern," some participants objected to requiring that all three of the characteristics recommended by the NSABB WG be met. Dr. Millett felt that the threshold for risk requiring additional oversight had been set too high and that any research that would be expected to produce a pathogen with any two of the characteristics should be considered for additional oversight. In this respect, Marc Lipsitch from Harvard University noted that the original GOF experiments that prompted the international controversy were initially believed to have met only two of these criteria, and eventually met only one. Thus, those experiments would not be subject to any of the oversight provisions under discussion. He and Dr. Inglesby argued that the third criterion was superfluous and that only issues of transmissibility and virulence need be considered. Dr. Lipsitch cited the original White House charge (White House, 2014a), public comments submitted to the NSABB by the Infectious Disease Society of America (IDSA, 2016), and "common sense" in support of his argument. John Steel from Emory University raised technical questions about how to measure these characteristics; for example, what does "highly" transmissible mean? He also cited the shortcomings of animal models to approximate transmissibility and called for additional guidance on how to make such decisions in practice.

The discussion also included a number of other specific reflections on the NSABB WG's Draft Working Paper and its recommendations

- Dr. Inglesby highlighted how important it was that any relevant regulations or other measures governing GOF studies of concern apply anywhere relevant research is being conducted, regardless of whether the laboratory receives federal funds or whether it is found in the public or private sectors.
- The value of the NSABB making its recommendations broad enough to fit GOF studies of concern with any pathogen, rather than just those covered by the moratorium, was noted by Dr. Millett. He also suggested that any characterization of GOF studies of concern should not be based upon taxonomy but instead focus on functional characteristics as contained in the draft definition.
- Dr. Selgelid from Monash University in Australia raised the possibilities of making oversight arrangements progress along a spectrum rather than being treated as binary. In such a model, a single risk threshold would not be established (above which research would be governed by specific oversight measures), but rather increasing levels of oversight would apply as the relative risk of the work increases. He also commented that, rather than first making a judgment about the scientific merit of a study and then assessing whether it raised GOF issues of concern, it might

be better to include considerations of risk at the earlier stage. If two studies show equal scientific merit and neither is considered of concern, then—other things being equal—would it be better to fund the less risky study if one cannot fund both?

- Questions as to the efficacy of the existing arrangement for addressing biosecurity information risks were raised by Dr. Inglesby and Dr. Millett, who encouraged further reflection on suitable oversight.
- Some participants, such as Dr. Millett, felt that bodies involved with assessing risks and benefits could not be housed within either the health or the security architecture but should be located inside a neutral agency.
- Questions about the interface between the proposed regulatory framework developed by the NSABB WG and existing arrangements for GOF experiments with specific agents, such as the one implemented by the National Institutes of Health (NIH), were raised by Nicholas Evans from the University of Pennsylvania. The ethics report also was discussed, with Dr. Evans suggesting that the scope of ethical issues related to GOF studies of concern was considerably broader than the scope of those included in the consideration of benefits in other areas, such as human subjects research. This, he suggested, would seem to increase the challenges in suitably reflecting potential benefits of GOF studies of concern.

Dr. Fineberg began the period of taking responses from the NSABB members by welcoming comments on any issue but said he hoped that they would reflect in particular on the "core question of what qualifies as being of concern." He noted that there had been a variety of viewpoints expressed about the necessity of meeting all three criteria, the implications of thresholds versus a spectrum, and the question he had raised earlier of whether the starting point could enough to make research meeting only one criterion "of concern." Dr. Kanabrocki responded by clarifying that the NSABB had not changed its thinking, as some had suggested, about the third criterion. The NSABB WG had recognized that there was some misinterpretation of the original language to limit the criteria to only resistance to countermeasures. Instead the intent is to capture the broader question of pandemic potential. Dr. Berns added that he emphasized "what's important is what you wind up with," that is, the potential for a pandemic and this led to the question of whether or not there were existing defense mechanisms. He also commented on the difficulty of predicting the consequences of research and the challenges in attempting to quantify such risks. Dr. Stanley commented that the issue was whether the research could create risk significantly above the existing risk for that pathogen. Again, this tied to the questions of natural immunity or countermeasures.

In response to other questions and comments, Drs. Stanley, Kanabrocki, and Berns acknowledged the importance of the issues that had been raised, and they commented that the NSABB had struggled, for example, with the question of whether to offer broad recommendations or the more specific guidance for which some participants were calling. Dr. Berns commented that, even more than the definition, he thought the NSABB WG had struggled with the level at which the decision would be made. Should the final decision be made inside the government or by an outside group, such as a FACA committee? Efficiency might suggest handling the decision inside the government, but the public interest in transparency would argue strongly for a FACA committee. They also noted that, although the NSABB was tasked to make recommendations for the U.S. government, they recognized the importance of the international dimensions of the issue. Dr. Stanley thought all the members of the NSABB, as well as its sponsors, believed that it was necessary to "strive for a global solution here, and that some type of harmonization essentially of these processes would be extraordinarily valuable." The process was definitely not completed and they welcomed the input the NSABB WG would receive during the 2 days of the symposium.

LESSONS FROM THE RISK AND BENEFIT ASSESSMENT

Charles Haas from Drexel University, a member of the Symposium Planning Committee, introduced the goals of the session. The details of the risk and benefit assessment (RBA) had already been reviewed in detail at the January 2016 NSABB meeting.[2] The purpose was therefore to build on those prior discussions to consider how risk assessment more broadly could serve the important roles that the NSABB's draft findings and recommendations, including its proposed conceptual approach for making decisions about GOF studies of concern, had given to judgments about risks and benefits.

Rocco Casagrande from Gryphon Scientific provided an overview of the RBA (Gryphon Scientific, 2015) as basic background for the session. The purpose of this 8-month study was to provide data on the risks and benefits associated with research on modified strains of influenza viruses

[2] The webcast of those discussions, along with copies of presentation slides and written public comments, are available at http://osp.od.nih.gov/office-biotechnology-activities/event/2016-01-07-130000-2016-01-08-220000/national-science-advisory-board-biosecurity-nsabb-meeting.

and the coronaviruses. The RBA had been divided into three major tasks, each of which required a distinct data collection and analysis approach: a quantitative biosafety risk assessment; a semi-quantitative biosecurity risk assessment; and a qualitative benefit assessment. Dr. Casagrande noted that the RBA was comparative; it determined the change in risk from research on GOF pathogens compared to research on wild type pathogens and identified the benefits to science, public health, and medicine afforded by GOF research compared to alternative research and innovations.

Dr. Casagrande presented key findings from the RBA. The biosafety risk assessment included a map of the series of events necessary for a laboratory incident to result in a pandemic. The probability of each event resulting in the next necessary event was determined. The RBA established that only a small minority of laboratory incidents with the most contagious influenza viruses would cause a local outbreak, and only a minority of those would lead to a global pandemic.

The published RBA had identified the pandemic strain of the 1918 H1N1 influenza virus as posing the greatest risk. However, subsequent information made available to Gryphon Scientific at the January NSABB meeting by Dr. Kanta Subbarao from NIH showing a high degree of cross protection afforded by exposure to the 2009 influenza against the 1918 influenza enabled a reassessment. Further analysis determined that the naturally circulating 1957 H2N2 influenza virus became the "riskiest" pandemic strain because its antigenic profiles would cause about 100 times more global cases, although it is only one-tenth as deadly as the 1918 strain.[3] As a result, it became the comparator against which other risks should be evaluated. The RBA also determined that the riskiest modified strain was a 1918 H1N1 strain altered to evade residual immunity or to be otherwise more transmissible.

Other key findings from the biosafety assessment included

- Manipulating GOF seasonal influenza strains at the BSL3 level may compensate for the increase in risk posed by the modified strains, largely because the extra system of respiratory protection decreases the risk of a laboratory acquired infection.
- Some of the manipulations that could theoretically increase risk may not be achievable or desirable. For example: (i) a strain that can overcome protective vaccination increases risk only if it can evade vaccine protection via immune modulation, not antigenic

[3] The details of the Gryphon Scientific analysis are available in supplemental material on its website: http://www.gryphonscientific.com/wp-content/uploads/2016/03/Supplemental-info%E2%80%93Protection-against-Infection-with-1918-H1N1-Pandemic-Strain.pdf. The final version of the report was released in April 2016 (Gryphon Scientific, 2016).

change; (ii) the scientific value of increasing the transmissibility of influenza virus beyond that of the most transmissible strains (or final titer beyond 1E8) is questionable and perhaps infeasible; (iii) there is no animal model of transmission for the coronaviruses, so manipulation of this trait is not currently achievable; and (iv) some estimates suggest that severe acute respiratory syndrome coronavirus (SARS-CoV) may already be more transmissible than was estimated in the RBA, in which case further manipulation would not affect risk.

The biosecurity risk assessment had two components: an assessment of the risks from acts targeting a laboratory; and security risks derived from the information generated by the studies. Key findings of the assessment of risks from acts targeting a laboratory included

- The traits that drive risk are similar when considering biosafety and biosecurity because the pathogens are transmissible. How the initial infections were caused is of little consequence once a local outbreak begins.
- Biosecurity events are often predicted to involve the covert infection of the public, so this type of an infection is much more likely to cause a global outbreak. By contrast, laboratory workers benefit from health surveillance and isolation protocols not available to the general public.
- To match the risk posed by biosafety incidents given a historical rate of laboratory acquired infections, a biosecurity event that covertly infects a member of the public must occur only once every 50-200 years. These events include theft of an infected animal, contaminated piece of equipment, or viral stock. Given the frequency with which these biosecurity events have happened, the RBA suggested that biosecurity be given as much consideration as biosafety.

The information biosecurity risk assessment analyzed "the risk that a malicious actor might misuse the information in publications describing GoF research" (Gryphon Scientific, 2015: 212). Key findings included

- Minimal information risk remains for GOF studies in influenza viruses because dual use methods have already been published.
- Significant information risk remains for GOF studies in the coronaviruses, but these studies are hampered by a lack of model systems.
- Information risk could easily be generated by research on other transmissible pathogens.

The benefits assessment identified GOF studies providing critical or unique benefits for both

- Influenza viruses, including studies that enhance viral growth from low titer; lead to evasion of residual or induced immunity; enhance virulence; enhance transmissibility; and lead to evasion of therapeutics in use and in development. And,
- Coronaviruses, including studies that alter host tropism; enhance virulence; and lead to evasion of therapeutics in development.

Dr. Casagrande highlighted a number of lessons learned during the execution of the RBA. He stated that the distinction between seasonal and pandemic flu is artificial because an old seasonal flu strain could become a new pandemic strain (as highlighted by 1957 H2N2 replacing 1918 H1N1 as the riskiest pandemic strain). He noted the lack of data on human reliability in life sciences laboratories in contrast to data from other well-researched sources such as the nuclear, chemical, and transportation industries. Those data show that human error is the most common cause of accidents. To use an example from the life sciences, it is more common for a lab worker not to use a powered air purifying respirator (PAPR) properly than for a PAPR to be defective. He also cited the difficulty posed by having no risk benchmark for work with wild type pathogens and the difficulty posed by restriction of the RBA to influenza and coronaviruses.

The RBA was then applied to a number of specific experiments, including those that

- Include virulence factors from 1918 H1N1 influenza in a 2009 H1N1 strain, which did not increase the probability that an outbreak escapes local control and indicated that global consequences scale linearly with case fatality rate.
- Aim to create antigenically distinct strains of a recently circulating seasonal influenza strain, which resulted in strains having a 2-3-fold increase in risk of escape, capable of inflicting 10 times more global deaths, resulting in a 20-30-fold increase in risk of infection. The meaning of this risk increase is difficult to interpret in the absence of standards for risk tolerance but suggests that more controls and measures should be taken to control infection risk from this modified pathogen than from the wild-type pathogen.

Dr. Casagrande also noted that bench researchers may not be familiar enough with the epidemiological properties of pathogens to properly characterize their strains. Guides or tools are needed to easily obtain parameter values for wild type strains and, perhaps, to aid with the calculations.

A series of commentators provided reflections on the RBA. They were asked to consider

- What they know needs to occur, based on their prior experience in the context of policy making, to make use of the Gryphon Scientific analysis and other information.
- The potential value of risk–benefit analysis in making decisions on individual cases of proposed research projects rather than the role of a study intended to cover an entire class (i.e., GOF) of investigation, which was the purpose of the Gryphon Scientific analysis.

Louis (Tony) Cox from Cox Associates highlighted the value of attempts to quantify risk in the RBA (Cox, 2016). Dr. Cox also discussed risk management, or what to do about that risk, especially as it related to determining which proposals to fund. Dr. Cox highlighted the value of clearly defined decision rules and conditional decision rules, detailing the conditions that would need to be met before a proposal might be funded. Dr. Cox reflected that efforts to determine the maximum acceptable risk were not useful approaches in a GOF setting. He argued that both the context and the benefits needed to be taken into account and suggested that attempting to improve the risk–benefit profile may be a more suitable approach. Dr. Cox suggested that "arbitrary coherence"—accepting risks because they are less risky than those we already accept—was also not appropriate in a GOF context. He believed that benchmarks and precedents were not necessarily the most appropriate basis for decision making but supported gathering more information before making funding decisions, including on opportunity costs. He asserted that there is a need to learn from past experience and to make the decision-making process adaptive. Dr. Cox also identified a series of specific proposals for strengthening funding decisions on GOF studies of concern (see Box 2-2).

Kara Morgan from Battelle Memorial Institute reminded participants of the difficulty of decision making on low-probability, high-risk events. Dr. Morgan introduced a number of tools developed to assist in such situations and help match decision-making complexity to potential risk. She discussed three frameworks for decision making, describing the frameworks as part of a continuum, enabling their adaptation to different contexts (see Box 2-3).

Dr. Morgan concluded that decision making is a social process, not an analytical one. There is a need for a process to help move from analysis to a decision. She advised the symposium that decision frameworks, rules, and process were just as important as the analysis.

BOX 2-2
Proposals for Improved Funding Decision Making on Gain-of-Function Research of Concern

- A decision rule maps available information to decisions—specifically whether to fund, not to fund, to require modifications before funding, or to seek additional information on which to base a funding decision.
- The performance of a decision rule can be evaluated for a stream of simulated projects with specified risk, cost, benefit, and information/uncertainty characteristics and proposer response characteristics.
- If we know enough about GOF research to simulate realistic project proposals and decision rule performance, then simulation-optimization of decision rules can lead to better (higher-performing) individual project-funding decisions.
- Otherwise, eliminate dominated decision rules (e.g., risk matrices, simple additive scoring systems).

SOURCE: Cox, 2016.

Adam Finkel from the University of Pennsylvania set out five factors to strengthen risk–benefit analysis that should be integrated into the development of the policy framework for GOF studies of concern (see Box 2-4).

Dr. Finkel noted considerable differences in opinion among different risk estimates of GOF studies of concern. He argued that risk was not a binary state and this provided the potential for a hierarchy of decision rules. He also noted the importance of including justice and equity for those individuals affected by risk. Dr. Finkel commented that it becomes much harder to assess risk when uncertainties exist and they are uncorrelated. He also felt that it was necessary to do a better job of communicating the benefits of GOF research. He called for further efforts to identify where the faults that lead to risk are occurring. He introduced a new study of existing best practices in regulatory excellence based on the concept of "listening, learning, and leading" developed through work in the Canadian energy sector (Coglianese, 2015).

Dr. Finkel discussed the importance of basic laboratory safety. He believed the best way to prevent accidents from infecting the population was to prevent them from infecting laboratory workers. Dr. Finkel concluded by encouraging the use of a more solution-focused risk–benefit analysis—where options are not restricted to a specific limited set of options—but one which focuses on the underlying policy need. He provided examples from sources of drinking water and synthetic biology. He cautioned that uncertainties rarely cancelled each other out in practice.

BOX 2-3
Different Models for Risk-Based Decision Making

Acceptable risk—In this model the risk is estimated along a spectrum. A boundary or threshold is set above which one decision would be taken and below which a different decision would be taken. The main challenge with this approach is derived from the innate uncertainty of science. While it is possible to mitigate this through the use of safety factors, it can still result with benefits not being taken into account. The process of determining where the boundary falls can also be challenging and past discussions on GOF have already demonstrated notable differences of opinion on this point.

Risk–benefit assessment—This is a two-factor analysis and builds on an understanding that societies are often prepared to tolerate some risk if they receive benefits in return. While this model does take benefits into account, to be fully effective, it is necessary to express both risks and benefits in comparable terms, preferably using the same metrics. This can often involve value judgments. A decision as to where the appropriate balance lies between risks and benefits is often subjective.

Deliberative criteria-based frameworks—This model allows the introduction of more factors. It enables the integration of different views through the use of criteria identified in advance of assessment. It can include both scientific contexts based upon observations and perceptions (such as facts, data, analytical results, assumptions, and uncertainties) and social contexts based on values (such as goals, objectives, priorities, concerns, ethical issues, non-observable criteria, policy decisions, and tradeoffs). This model is more resource intensive than the other approaches but is more collaborative.

SOURCE: Morgan, 2016.

BOX 2-4
Factors for Strengthening Risk–Benefit Analysis

1. Risk and benefit estimates should be balanced, quantitative, humble, explicit about value judgments, and channeled in service of a thoughtful decision rule.
2. Benefit estimates can be made commensurable with risk estimates, and should be communicated with equal care.
3. Purely risk-based prioritization is inferior to net-benefit prioritization.
4. Transparency in public engagement is important, but not as important as "apparency" (which provides information on rationale and motivations).
5. "Solution-focused" analysis of GOF and public choice may require wholly new institutional arrangements, not just incorporation into existing policy frameworks.

SOURCE: Finkel, 2016.

Discussion

The resulting discussion began by highlighting the importance of having good baseline data against which to measure risk: for example, through a national database or framework of laboratory near misses, accidents, or disclosures, as discussed by James Welch from the Elizabeth R. Griffin Foundation. Panelists noted that the U.S. government had already committed to develop such a database (U.S. Government, 2015: 4).[4] The need for additional resources to undertake focused research to fill data gaps was highlighted by Gigi Kwik Gronvall from the UPMC Center for Health Security. Shortages of data on benefits and risks were felt by several participants to apply to infectious disease research and emerging areas of life science research more broadly. The importance of tools to enable scientists to operate safely and securely on an ongoing basis was also noted by Dr. Gronvall. There was also a discussion, prompted by Allison Mistry from Gryphon Scientific, of the need to differentiate between conducting functional changes in wild type as opposed to research backbones or chassis. The value of including comparative risks in different chassis in definitions of GOF and GOF studies of concern was also explored.

Participants also considered the limits of comparing risks and benefits in this type of research. The discussion explored the challenges in suitably reflecting the potential public health benefits of research. Corey Meyer from Gryphon Scientific, who had led the benefits assessment portion of the RBA, made a number of comments. She said that although it may be possible, at least qualitatively, to compare the risks and benefits of research for public health, she was not sure that was true for the benefits of scientific knowledge. She also wanted to underscore that "while the risks of the research are immediate in that they are occurring at the time the research is being conducted, the benefits to public health will be realized in the future. And there is significant uncertainty in how long it will take for those benefits to be realized because translation of basic science research into public health benefits is complex and depends on many other factors."

Adam Finkel commented that there is a substantial literature on discounting and the time value of benefits on which one could draw. He thought the problem was not intractable and offered the example of climate change research, where he said there is a movement toward lower discount rates. In this area, "the future speaks more loudly than we

---

[4] The recommendation—"Establish a new voluntary, anonymous, non-punitive incident-reporting system for research laboratories that would ensure the protection of sensitive and private information, as necessary"—is one of the products of the Federal Experts Security Advisory Panel, whose report was made public in October 2015 (U.S. Government, 2015).

have allowed in the past." So that part is not at all intractable. He cited the example from Michael Selgelid's white paper of benefits in terms of expected lives saved.

Rocco Casagrande commented that the daunting part of assessing future benefits is not how much to discount potential lives saved but how to make an estimate of how likely it is that any scientific discovery will lead to a public health benefit. Tony Cox commented that it also depends on what other research is done. And Kara Morgan noted that even failed experiments may offer useful lessons. John Steel also noted that such research can help ameliorate disease events that happen infrequently but that potentially result in tens of millions of deaths.

Some participants, such as Marc Lipsitch, questioned the findings on the unique benefits of certain GOF research, suggesting that the knowledge could have been generated using alternative approaches. In his view, the net contribution of GOF research to knowledge on influenza viruses has been overstated. He also said that the knowledge about mutations and phenotypes identified by Fouchier and Kawaoka had already been identified in previous safe experiments. The confirmation that they were important in the GOF context was new, but he asserted that their utility for public health prediction was so far unproven. "So the net benefit for public health is much smaller than the net knowledge." Issues around identifying and ensuring sufficient oversight of dual use research more broadly were also discussed, including that, as life science and biotechnology tools are getting more powerful, the potential for their misuse for malign purposes might also increase. On the RBA, some participants, such as Piers Millett, reflected that the process of updating the risk comparator from the 1918 influenza strain to the H2N2 1957 pandemic strain was a practical example of the importance of the inclusion of the concepts of innate or acquired immunity against pathogens in the third set of characteristics proposed by the NSABB to define GOF studies of concern. He also suggested that the RBA was a missed opportunity to explore the international opportunity costs associated with different decisions on GOF studies of concern, from a moratorium on relevant research, through increased oversight, to taking no additional steps.

The shortcoming of existing arrangements in identifying and mitigating biosecurity information risks was noted by some participants, including Victoriya Krakovna from the Future of Life Institute, Dr. Millett, and Megan Palmer. They argued that these assessed risks were only low because the critical information had already been released into the public domain over the last decade. This led to questions about the efficacy of current arrangements to identify potential future biosecurity information risks, such as those for coronaviruses highlighted in the RBA. The value of encouraging comments and reflections on the RBA and associated methodologies from a wider group including different types of expertise was also noted by Megan Palmer.

THE SCIENCE OF SAFETY AND THE SCIENCE OF PUBLIC CONSULTATION

Baruch Fischhoff from Carnegie Mellon University, a member of the Symposium Planning Committee, opened the session by explaining that there had been a successful session at the first Academies GOF symposium, which offered an introduction to the lessons from research into human factors, public consultation, and risk assessment to inform the preparations by NIH and the NSABB for the RBA. This year the planning committee had organized another session focused on the insights that social science research can offer about the design and implementation of federal oversight for GOF studies of concern. The panel included experts in organizational culture, human factors, and public consultation who would offer comments on the NSABB draft recommendations and specific suggestions for the ultimate choices to be made by the U.S. government.

Ruthanne Huising of McGill University introduced the insights about compliance with safety regulations in life science laboratories gained from past research in which she had taken part (see, e.g., Huising and Silbey, 2011). Since 2012 she had also been observing Canadian regulators design new biosafety and security regulations that went into force in December 2015 using an intensive public consultation process.

Dr. Huising discussed behavior and decision making as mediated by social organizations, which can include both formal social structures (such as organizations and families) and what she termed emergent systems of meaning ("culture"), which include norms, values, and assumptions. The incidents that led to the GOF deliberative process had provoked extensive discussions of the existing culture of life sciences laboratories as this affected safety and security. In these and similar discussions, the concept of culture is often treated as both the "problem" (a "lax culture" or "insufficient culture") and the "solution" ("build a culture of safety," "change the culture"). Culture, she argued, is often understood as a managerial tool. She described how concepts of culture can be applied to understand how laboratories approach and implement safety provisions.

Dr. Huising described how culture might be shaped through socialization processes. Beginning with graduate training, researchers are observing and learning how successful members of their field think, talk, and act. They learn how competent, respected members of the community behave, potentially through their attention to safety, security, and risk.

Safety cultures can be designed and Dr. Huising provided examples of the systems used by BP and Dow Chemicals. Such efforts tend to be top down and centralized, she noted. They can be slow to develop and expensive, and they often ignore differences in interests and resources. She suggested that safety cultures can also emerge, resulting in multiple, heterogeneous cultures. Such change often occurs in response to shocks, with new values and norms emerging. This approach can be slow, but it is self-reproducing. Dr. Huising felt this model might be more suitable for the scientific endeavor, in part because it would better reflect the nature of the organizational structure of research laboratories.

The organizational structure of relevant institutions can also impact culture, Dr. Huising noted, with administrative and academic laboratory components operating with different logics. Academic administration is organized in ways that give it considerable similarities to the organization of regulatory and other government agencies. In contrast, the laboratories, at least in theory, operate through collegial governance and a democratic approach to organizing. That said, principal investigators (PIs) have remarkable autonomy in how they organize and run their laboratories. Dr. Huising commented that "Decision making is highly decentralized, and often operates according to verbal agreements. Trust is a very important component in how things get done in these laboratories." Because the laboratories work on soft money, they are often in flux, with continuously changing resources, members, and activities. She also noted that these different professional bureaucracies have implications for biosafety and biosecurity, in particular by determining responsibility for legal and administrative requirements, allocating, authority to enforce those requirements, and facilitating compliance.

Dr. Huising presented key findings from studies of safety culture in biology laboratories. She emphasized that these studies came from BSL2 facilities because of the difficulties in obtaining sustained access to higher containment (BSL3, BSL4) facilities. The findings include:

- Researchers experience compliance requests as intrusions and impediments to their work. They communicate safety as peripheral to their research work and sometimes delegate it to students. They are most likely to incorporate safety features into their practices when they align with efforts to control physical materials.
- Most violations are minor (housekeeping). A small number of laboratories account for the majority of violations.
- Organizations depend on environmental, health, and safety (EHS) staff (such as Biosafety Officers) to ensure compliance.

With regard to the last finding, she noted that the roles of the EHS staff included buffering researchers from record-keeping, inspections, corrections, and helping to maintain compliance. They negotiate increased daily compliance by working in laboratories, generating familiarity, trust, and relationships, which also gave them the ability to anticipate problems and to identify emerging dangers. In many cases, the EHS staff was able to draw on requirements and regulations to increase their resources and authority in relation to faculty, but she commented that these "boots on the ground" were chronically underfunded.

She noted the emergence of a "responsibility movement" in other facets of the life sciences, with examples of good practice in green chemistry, nanotechnology, synthetic biology, and the citizen science movement. Dr. Huising concluded by providing a number of specific recommendations for developing policy options for GOF research (see Box 2-5).

Gavin Huntley-Fenner from Huntley-Fenner Advisors introduced concepts in human factor research relevant to the NSABB WG's Draft Working Paper and recommendations. Dr. Huntley-Fenner highlighted that, in general, human error has increased proportionally as a contributor to accidents. He recalled that for laboratory biosafety, despite advances in technology, instruments, and personal protective equipment, the World Health Organization had asserted that "human error remains one of the most important factors at the origin of accidents" (WHO, 2006). He noted that there was a lack of data on human reliability in laboratories, and he stressed the importance of collecting more data on safety. He cited the conclusion in the RBA that "The state of knowledge of the rates and consequences of human errors in life science laboratories is too poor to develop robust predictions of the absolute frequency with which laboratory accidents will lead to laboratory acquired infections" (Gryphon Scientific, 2015: 3) to underscore the relevance for GOF policy deliberations.

BOX 2-5
Shaping Cultural Change Relevant to the Oversight of Gain-of-Function Research of Concern

- Culture change should come from within the scientific professions, making it more likely to produce long-term, global changes.
- Particular focus should be placed on the roles of Biosafety Officers in relevant laboratories, and that will require resources and support.
- Research about daily decisions and practices in laboratories needs to be supported and expanded to encompass higher containment facilities, providing for better baseline data.

SOURCE: Huising, 2016.

There are a range of factors that can contribute to the emergence of error, including how physical and cognitive stresses undermine human reliability, according to Dr. Huntley-Fenner. He suggested that the comparative scarcity of accidents might still mask latent risks, with more numerous incidents and errors going unreported. He highlighted research by the Government Accountability Office in 2009 that concluded latent risks still exist in laboratories from underappreciated human error (GAO, 2009). He suggested that human factors research could provide tools for designing, implementing and maintaining systems in which errors are mitigated when they occur. The benefits of incorporating human factor principles were potentially significant, with Dr. Huntley-Fenner suggesting that they could reduce risk associated with GOF studies of concern substantially. He noted that some simple approaches could yield a significant reduction of errors, such as the development and use of simple checklists, which had a significant impact in reducing surgical errors in hospitals in both developed and resource poor countries (Haynes et al., 2009).

Progress has been made in other areas to address shortcomings in human factor safety data. For example, Dr. Huntley-Fenner discussed how the National Aeronautics and Space Administration has succeeded in mining data it already had in ways that provided insights into areas where it had less data, which was then used successfully to reduce risk (Chandler et al., 2010). He argued that limited relevant laboratory safety data do exist. For example, a survey of laboratory acquired infections in 68 institutions in Belgium indicated that 95 percent of the incidents involved human error (Willemarck et al., 2012: 14). He suggested that the human factors research community was well positioned to provide relevant data but more work was needed in high-containment laboratories.

Measuring incidents was only one necessary step; controlling incidents was also important, noted Dr. Huntley-Fenner. He highlighted research that showed the success of applying multifaceted controls. He also highlighted the importance of considering context. Guidance from the United Kingdom on human factors that result in noncompliance with standard operating procedures demonstrated that cutting corners was mainly "due to situational and organizational factors. These factors include, for example, time pressure, workload, staffing levels, training, supervision, and availability of resources" (Bates and Holroyd, 2012). Dr. Huntley-Fenner recommended rigorous data collection and sophisticated analytics to reduce risk associated with GOF studies of concern (see Box 2-6). The self-driving Google car was provided as an example of successfully gathering and leveraging data on human decision making and error to build a system that reduces those risks.

Monica Schoch-Spana from the UPMC Center for Health Security outlined four basic considerations for the design of public deliberations.

BOX 2-6
Improving Rigorous Data Collection and Sophisticated Analytics to Reduce Risk Associated with Gain-of-Function Research of Concern

- Create national reporting standards that go beyond the most significant adverse events
- Collect data on near misses
- Collect data across multiple bodies to counteract relative rarity of events
- Standardize data inputs whenever possible
- Develop analytics driven models of when and what adverse events are more likely to occur and under what circumstances
- Direct training and other interventions where they are needed most

SOURCE: Huntley-Fenner, 2016.

1. **Which public(s) to involve in deliberations**—In the context of GOF studies of concern, Dr. Schoch-Spana suggested that considering three overlapping categories would be useful: the *pure public*, or naive citizens; the *affected public*, or persons or groups whose lives are altered or influenced by a policy decision; and the *partisan public*, or representatives of groups with vested interests or expertise in the policy matter. She also noted that each of these categories of the public had been engaged in past discussions on GOF studies of concern, with the affected public implied in the RBA, affected publics and the pure public noted in the ethics analysis, and partisan publics reflected in relevant publications and comments.
2. **What is the purpose for public(s) deliberation**—Three distinct aims were highlighted: *knowledge exchange*, conveying information from policy makers to publics, or transmitting views, opinions, or attitudes from the publics to policy makers; *innovation*, eliciting rich unpredictable insights that come from crowd-sourcing a problem or from experiential, on-the-ground knowledge; and *democratic accountability*, ensuring broad representation in a decision about the common good. If the public deliberation on GOF studies of concern was intended for democratic accountability, Dr. Schoch-Spana noted it was necessary to give people the time, information, space, and authority they need to perform that role. Merely bringing "ordinary people" or a cross-section of society together to deliberate does not automatically achieve this aim.

She suggested a series of desirable characteristics for public deliberations on GOF studies of concern, including diversity, balance, civility, accountability, and consent.

3. **Which process enables the public to fulfill its purpose**—The use of three types of processes in the GOF deliberative process were reviewed: *communication*, a form of transparency through putting out information for the public—for example, press releases, educational websites, and summary reports such as those made available by the first Academies GOF symposium and the NSABB GOF meetings, as well as making the RBA available online; *consultation*, a means of gathering input, such as through enabling public comments on draft NSABB recommendations and to the U.S. government on future funding and oversight policy; and *collaboration*, a more deliberative option to exchange ideas and share responsibility for making and implementing policy. To date, she felt that the life sciences and other partisan publics have had strong input but deliberation with the broad public has not yet been explored.
4. **On what problem will the public(s) deliberate**—Dr. Schoch-Spana reviewed good practices in identifying problems, especially where there are conflicting values as to the public good, and for controversial and divisive topics. She used them to identify three questions on which the publics might deliberate for GOF studies of concern: (i) Despite potential contributions to public health, should studies that could produce a pathogen of pandemic potential be performed at all?; (ii) Are finite dollars better spent on experiments to create pathogens with pandemic potential (which produce unique knowledge) or on strengthening the rest of the flu preparedness portfolio?; and (iii) If any, what added steps should trustee institutions (e.g., the U.S. government or research entities) take to strengthen pathogen of pandemic potential biosafety and biosecurity protections and public confidence in them?

Dr. Schoch-Spana also discussed how to operationalize standard elements of deliberation design. She noted that there is no single methodology for public deliberation, but she did describe a number of minimum standards for public deliberation, in particular for inclusivity and diversity, the provision of information, value-based reasoning. She also discussed methods for measuring the success of the process. Dr. Schoch-Spana concluded that meetings to date have engaged individuals from the life sciences, security, public health, biosafety, risk analysis, and the drug and vaccine industries, but the general public had been largely absent. She identified an unresolved issue of whether more sophisticated, resource-intensive deliberative sessions could be held outside the present circle of vested parties. A number of possible activities for such a process were suggested (see Box 2-7).

In opening the floor for discussion, Baruch Fischhoff commented that the panelists had been encouraged to offer recommendations based on their own professional experience and research. He added that for those in the audience who were not familiar with the social, behavioral, and decision sciences, the panel should have provided some idea of the breadth and depth of the research that is available if one wanted to put the human aspect of this enterprise on a scientific foundation. It also illustrated the mix of methods used in this research: various theories; multiple methods of observation, including direct observation and laboratory and field experiments; traditional and statistical and analysis; and various types of data.

Discussion

The resulting discussion further elucidated specific aspects of the presentations. The importance of additional data gathering on accidents and associated human factors research was a repeated theme. Susan Wolf, an NSABB member from the University of Minnesota speaking in her personal capacity, raised operational issues around data collection, data standards, and the development of data collection systems. Gavin Huntley-Fenner commented that the dearth of current data on accidents and human reliability in laboratories does not mean that what people want to know is unknowable. He and others also noted the value as well as the potential challenges in implementing confidential accident reporting. The need to ensure that comprehensive reporting systems for

BOX 2-7
Examples of More Sophisticated, Resource-Intensive Deliberative Activities to Extend the Current Process

- Initiate a formal evaluation process to determine how the (primarily) partisan publics rate the quality of deliberations in terms of inclusivity, information provision, and value-based reasoning.
- Hold deliberative exercises in communities now hosting facilities where GOF studies of concern are undertaken.
- Engage a cross-section of the American public in a deliberative exercise about a specific question.

SOURCE: Schoch-Spana, 2016.

human errors are developed and implemented in a nonpunitive manner was stressed by Kavita Berger from Gryphon Scientific. In response, Dr. Huntley-Fenner noted the importance of even seemingly small things, such as finding language for reporting forms that did not use negative categories ("theft," "loss"), and designing systems so that there was feedback or other incentives for reporting, such as providing information that could be used to improve safety. Ruthanne Huising said that the new regulatory framework developed in Canada had a nonpunitive reporting system that offered potential lessons about dealing with privacy issues and offering useful feedback.

Given what he saw as the difficulties of implementing a nonpunitive system in high-containment laboratories, Andrew Kilianski, a National Research Council Fellow from the Edgewood Chemical Biological Center Aberdeen Proving Ground, made the specific suggestion to conduct research focused on the possible relationships between human error by graduate students under minimal-containment standards and other indicators of their proficiency. Megan Palmer from Stanford University highlighted the importance of strategic interventions to allow sustained scholarship on the social and behavioral dimensions of research. Monica Schoch-Spana commented that best practices for biosafety and biosecurity have not been captured, synthesized, and disseminated by researchers.

Adam Finkel said he was concerned that there had not been a discussion of a confidential channel for reporting incidents, citing what he thought was becoming a less favorable climate for "whistle-blowers" in many settings. He also stressed the need to consider outside incentives to support a culture, including enforcement. He thought that traditional regulation was probably not appropriate but cited other models, such as third-party audits, that could be considered.

Issues around the enforcement of safety and security regimes were also addressed, with some participants noting that a subsection of accidents and incidents are a result of negligence and malfeasance, requiring some form of censure. These individuals highlighted the importance of access to necessary resources for enforcement.

Opportunities for strengthening safety by designing out the consequences for human error were also noted: for example, by Dr. Finkel. He cited useful precedents from health care settings and commented that he sensed opportunities were not being widely studied or implemented in laboratory settings. In response, Dr. Huntley-Fenner noted a paradox that, as one designs out the other sources of error, human factors become an increasing portion of whatever error remains. That is not a reason to neglect those helpful improvements, but it is a reminder that human error will always be with us.

Kavita Berger noted past work on behavioral threat assessments and asked whether there were methods that could be applied earlier to detect individuals who posed potential biosecurity threats such as, for example, someone stealing an agent or animals, vandalism, violence, or deliberate misuse.

Participants discussed a multilayered approach as raised by Dr. Schoch-Spana, highlighting the need to ensure that such a system includes public engagement and transparency at all levels. Susan Wolf asked about the potential value of having a formal FACA committee for GOF studies of concern. Dr. Schoch-Spana commented that a FACA committee would satisfy one level of engagement and could be beneficial, but one should think of shared governance across all levels. She cited the systems in place at Duke University and St. Jude Children's Research Hospital (see next chapter) as examples worth studying for approaches to providing a diversity of views and participants. She commented on the need for more efforts to collect and share best practices about ways to improve biosafety, biosecurity, and what she called "bio-credibility." The potential additional burden imposed on scientists involved with GOF studies of concern from participating in further public deliberation exercises was raised by Margaret Kosal from the Georgia Institute of Technology. She asked if this was another unfunded mandate to educate a sometimes ignorant public that might be hostile to science for reasons that have not come up in these discussions.

David Drew from the Woodrow Wilson Center introduced himself as a "concerned citizen" who had not been familiar with the GOF controversy before the symposium. He raised the issue of whether the type of public engagement by scientists represented at the symposium was actually a form of "upstream engagement," which can be interpreted as designed to defuse the public's concerns without really addressing them. Dr. Schoch-Spana responded that to be effective the public deliberation process needs to be a shared dialogue that leads to mutually agreeable common ground, not just pure persuasion. Silja Vöneky from the University of Freiburg said she appreciated the stress on the value of ensuring that culture change comes from within scientific disciplines. But she also noted other strong incentives for scientists, such as publication, and suggested broader consideration of opportunities to nudge scientists to strengthen their focus on the safety and security implications of their work. Dr. Huising commented that the issue of culture change is sensitive when one is dealing with elites. In this case one was dealing with highly educated elites who are used to substantial autonomy and are not necessarily very open to ideas that are coming from elsewhere. She believed strongly that the ideas about the importance of safety and security in science are going to have to come from some of the best researchers in each discipline. “We need the leaders in these disciplines to model the importance of these values and normative expectations in research. We need the journals to expect it and conferences to highlight it. It will have to be pushed from within to be effective.”

3

Issues for U.S. Policy

THE POLICY LANDSCAPE IN THE UNITED STATES

Michelle Mello from Stanford University, a member of the Symposium Planning Committee, introduced the theme of the session. Key Findings 2 and 3 of the National Science Advisory Board for Biosecurity (NSABB) Working Paper (NSABB, 2015a: 3-4) addressed the adequacy of the policy frameworks in the United States to provide oversight of gain-of-function (GOF) studies of concern. Finding 2 indicates that the frameworks are effective overall, yet Finding 3 suggests that their adequacy for managing the risks associated with GOF research may vary, depending on which pathogen is being studied.

She commented that there seemed to be plenty to discuss about where the policy frameworks may and may not be adequate or optimal for addressing these risks. To that end, the speakers were asked to reflect, depending on their institutions, on the issues facing federal agencies in administering this regulatory framework as well as some of the strengths and weaknesses in the current policy framework and opportunities for optimizing oversight of this area of research.

Gerald Epstein from the Department of Homeland Security reviewed the scope of the NSABB proposal in terms of who is covered, which pathogens and activities are covered, and what is required. In terms of the existing policy context, he wanted to differentiate between those that are in effect by force of law, and therefore affect all researchers in the United States, as opposed to those that are, for example, a condition of government funding, which would directly affect only the recipients of that funding. There may be indirect effects in other areas, but a funding hook would only directly affect recipients of U.S. government funding.

The first part of the existing regulatory context that affects everyone in the United States is the laws in place to prohibit biological weapons development.[1] This statute is the mechanism by which the United States implements the Biological and Toxin Weapons Convention, an international treaty which prohibits development or acquisition of biological weapons. Unfortunately, the law does not contain definitions of prohibited types of activity or agents, so he thought the level of subjective judgment involved in proving a violation makes prosecution difficult.

Partly for that reason, the Select Agent Regulations were developed and expanded through a series of statutes.[2] This is a comprehensive set of safety and security requirements governing any use of certain listed pathogens. Of the GOF pathogens, three of them—1918 flu, highly pathogenic avian influenza, and severe acute respiratory syndrome (SARS)—fall under these regulations. The Middle East respiratory syndrome (MERS) does not.

Under the Select Agent Regulations, institutions have to be registered, researchers and staff have to be vetted by the government, and the institution has to have permission to use those agents. There are requirements for safety and security and for incident response and reporting associated with use of these pathogens. And in this case, the government does not have to prove intent. If one is found with one of these agents and has not registered with the government, it is a violation of a law. When that law was passed, it was also recognized that there are legitimate and important reasons why these agents need to be used in research. This is why there is a process by which research institutions and people can become vetted and approved to work with these agents. But it does provide a barrier for people who are not within that scheme.

A third area of legislation that binds everyone in the United States is export controls. These affect the export of certain listed pathogens from the United States or the communication of certain nonpublic, proprietary information that could, for example, include information about how to develop a particular strain of a pathogen if that were not published in the open literature. Information that is published in the course of fundamen-

---

[1] The primary statute to implement U.S. obligations under the Biological and Toxin Weapons Convention is the Biological Weapons Anti-Terrorism Act of 1989 (Public Law 101-298, May 22, 1990).

[2] The Select Agent program was created by the Antiterrorism and Effective Death Penalty Act of 1996 (Public Law 104-132, April 24, 1996). Following the attacks of September 11, 2001, and anthrax mailings, the program was expanded by the USA PATRIOT Act of 2001 (Public Law 107-56, October 26, 2001) and the Public Health Security and Bioterrorism Preparedness and Response Act, known as the Bioterrorism Act of 2002 (Public Law 107-188, June 12, 2002).

tal research is not affected by export controls, but there is a set of statutes and regulations that could have some bearing on the ability to do and disseminate biological research.

Because they had already been discussed, Dr. Epstein touched only on the policies that are attached to government funding. This includes the federal and institutional policies for oversight of dual use research of concern (DURC), which among the GOF pathogens covers only 1918 flu and H5N1 highly pathogenic avian influenza (White House, 2012, 2015b).

He commented that another framework developed by the Department of Health and Human Services (HHS) for certain H5N1 and H7N9 strains was very similar to the structure of the NSABB's recommendations (HHS, 2013). Lawrence Kerr would describe the framework during his presentation.

Dr. Epstein cited *Biosafety in Microbiological and Biomedical Laboratories* (CDC and NIH, 2007) and *Guidelines for Research Involving Recombinant or Synthetic Nucleic Acid Molecules* (NIH, 2013), extensive biosafety and biosecurity guidance that the Centers for Disease Control and Prevention (CDC) and the National Institutes of Health (NIH) developed for use by anyone doing biological research. It is obligatory as a condition of NIH funding for institutions to follow these processes. This set of best practices enables research using potentially dangerous pathogens to be done safely both for those working in the laboratory and for those in the community. Beyond the formal requirements, this guidance is used widely around the United States, not only for government-funded work, and indeed around the world. Even a policy that nominally has only the force of its ties to government funding can have much greater influence.

Finally, with the caveat that he was not a lawyer, Dr. Epstein cited the issues of the liability that any institution working with potentially hazardous substances could face. Any entity working on something that could pose a risk to its workers, to the neighborhood, or to the environment has to do so in recognition that if there were an accident that causes damage in the community they could be held financially liable. This includes not only harm to the institutions or employees but also harm to the general public. And the extent to which an institution could be held liable may depend on the degree to which there is a regulatory structure in place and whether the institution had been complying with those regulations.

Dr. Epstein commented that any additional development of policy related to GOF research would have to be embedded in the already existing frameworks, and the question of whether these existing procedures would have to be modified to fit the new one or whether they would sit on top would have to be determined as the policy process went ahead.

Dr. Kerr from HHS provided an overview of HHS framework for research with certain highly pathogenic avian influenza viruses (HHS,

2012). Dr. Kerr noted that during the research life cycle, there are points at which biosecurity concerns could be addressed, but it is too difficult—and too damaging to the research enterprise—to do this at the publication stage. Therefore, HHS focused on the research proposal stage as part of the funding award process. He noted that within the existing HHS GOF policy, the focus is on studies that could produce an agent with increased pathogenicity or transmissibility via respiratory droplets. In such cases, an extra level of review is required. The results of the review determine whether a proposal goes on to departmental level review. Seven criteria are taken into account during these reviews (see Box 3-1).

The department-level review provides multidisciplinary expertise—including public health, scientific, security, intelligence, countermeasures, and preparedness and response—from a number of agencies to evaluate these proposals. The department-level review will also identify any additional risk mitigation measures that should be required, and it will determine whether a given proposal is acceptable for HHS funding. For proposals that are deemed acceptable, the funding agency within HHS will make the final funding decision. Dr. Kerr indicated that only a small number of research proposals had undergone a departmental review, but he commented that the results reflected the full spectrum of what one might expect from a review process if it is working well. Some proposals received full approval by the committee and were recommended for funding to the funding agency director. There were also research proposals

BOX 3-1
Criteria Used to Determine If Research Is Relevant to the HHS Framework for Certain H5N1 and H7N9 Influenza GOF Studies

1. Such a virus could be produced through a natural evolutionary process;
2. The research addresses a scientific question with high significance to public health;
3. There are no feasible alternative methods to address the same scientific question in a manner that poses less risk than does the proposed approach;
4. Biosafety risks to laboratory workers and the public can be sufficiently mitigated and managed;
5. Biosecurity risks can be sufficiently mitigated and managed;
6. The research information is anticipated to be broadly shared in order to realize its potential benefits to global health; and
7. The research will be supported through funding mechanisms that facilitate appropriate oversight of the conduct and communication of the research.

SOURCE: Kerr, 2016.

that were received in which individual experiments were rejected by the committee and it was recommended to the funding agency that those not be funded.

Richard Frothingham from Duke University provided an overview of the review process for dual use research instigated by their Institutional Biosafety Committee (IBC), which has been reviewing research for dual-use potential since 2003. Its experiences were recounted in an article in *Science* in 2007 (Davidson et al., 2007). The committee determined that most projects with significant dual use potential were GOF studies, and as a result, it added seven questions to its recombinant DNA registration form in 2005. It also undertook specific training for IBC members on dual use research. The IBC has examined all research including recombinant DNA, select agents, and all research under BSL3 conditions as well as other research upon request.

The Duke IBC does not use a specific dual use definition or threshold but has identified relevant research through the NIH study section or program officer, by the principal investigator's (PI's) answers on the recombinant DNA registration form, or by members of the IBC during the course of its research reviews. Specific examples of GOF identified by the Duke IBC were provided, including cytokine expression by Ectromelia; virulence factors in uropathogenic *E. coli*; adaptation of dengue virus for growth in *Drosophila* cell lines; and HIV infectious molecular clone pseudotyped with vesicular stomatitis virus-G (VSV-G) for initial entry into renal cells. The Duke IBC had learned a number of lessons from having reviewed this research, including:

1. GOF studies were encountered regularly as part of the broad category of dual use research, but the IBC had yet to encounter GOF studies of concern.
2. PIs have had challenges with the concept of dual use research; it was possible to reach consensus on the dual use potential of most biomedical research, but not on specific categories (e.g., DURC).
3. Focusing on specific risk mitigation strategies rather than whether a particular experiment was DURC did enable the IBC to reach consensus, and no GOF research proposals have been rejected.
4. The Duke IBC had received external expert advice on some studies as part of the review process and this had been helpful.

Dr. Frothingham noted that the comparative scarcity of events involving the misuse of research to cause harm makes it difficult to measure the benefits from dual use reviews. He did highlight their value in building public trust in responsible science. He suggested that the early review of GOF research might reduce wasted effort by scientists and improve peer review and funding outcomes. He concluded by providing a number of perspectives on the NSABB recommendations (see Box 3-2).

Philip Potter from St. Jude Children's Research Hospital introduced the hospital's work on influenza, including its status as one of the National Institute of Allergy and Infectious Diseases' Centers of Excellence for Influenza Research and Surveillance and a World Health Organization Collaborating Center for Studies on the Ecology of Influenza in Animals. Dr. Potter noted that, as a result of their work with influenza viruses, St. Jude is likely to be affected by decisions over the oversight of GOF studies of concern. He highlighted the existence of a specific DURC committee that consists of both scientists and nonscientists. In their system, the PI is responsible for presenting the risks and benefits of the proposed studies. To assist the committee, St. Jude has developed internal guidance on what they should consider. This included ensuring that no GOF virus is resistant to antiviral agents, that suitable vaccines are available, and that advice about the challenges of evaluating risks and benefits in "gray" areas, such as research altering host range and or tropism, is available. The committee has also embraced the ferret as the gold standard for biological testing, requiring its use in all relevant experiments.

The DURC committee has also subjected all experiments involving H7N9 influenza virus to the same scrutiny as the 15 agents covered by DURC requirements. Dr. Potter noted that the DURC committee does

BOX 3-2
Perspectives on the NSABB Recommendations Drawn from the Experiences of the Duke Institutional Biosafety Committee

- GOF is easier to understand than dual use.
- The proposed definition of GOF studies of concern is much clearer than the current DURC definition. It should be possible to reach consensus in determining when GOF research is a GOF study of concern.
- The GOF studies of concern world should be small and definable. There will be substantial overlap with Select Agent programs. Institutional experience with Select Agents will be valuable in implementing GOF review.
- The current IBC or Institutional Review Entity (IRE) mechanisms seem appropriate for institutional GOF review. Institutions should have a low threshold for requesting external expert advice.
- Duke recently moved dual use review, including GOF review, out of the public IBC space to a confidential IRE. The process of GOF review should be transparent but the content is often inappropriate for public disclosure.

SOURCE: Frothingham, 2016.

not publish minutes. He also provided a number of perspectives on the NSABB draft recommendations and broader policy frameworks (see Box 3-3).

Discussion

The discussion that followed addressed local adaptation of the research covered in assessment of DURC or GOF studies of concern. J. Patrick Fitch from the Battelle National Biodefense Institute, a member of the NSABB, speaking in his personal capacity, raised the question of who would be responsible if something went wrong—the scientist or the committee? He also commented that his institutional committee had an experience similar to Duke's. In that situation, a focus on developing appropriate risk mitigation plans for relevant research, rather than on identifying a specific experiment as "DURC," had proved to be a much more productive approach to achieving the same goal.

Following a suggestion by planning committee member Barry Bloom from Harvard University, speaking in his personal capacity, participants also explored whether there was a need for separate IBCs, DURC committees, and possibly GOF studies of concern committees. Some participants felt that it might be possible to combine committees, especially if there was access to the additional expertise that might be required for new roles. Allison Mistry from Gryphon Scientific also proposed updating the DURC requirements to reflect GOF studies of concern and to

BOX 3-3
Perspectives on the NSABB Draft Recommendations and Broader Policy Frameworks

- This is a good initial draft that provides guidance to PIs and Institutional officials.
- The criteria for assessing GOF research are reasonable, but are not specific (terms "highly," "significant," and "likely" should be better defined).
- DURC Committees would likely have the expertise to assess GOF research.
- It is unclear whether local IBC and DURC committees can add additional science as DURC or GOF—if so, this might this lead to a patchwork of institution-dependent rules.
- GOF guidelines need to be crystal clear.
- Need to be specific about who to contact if issues arises.
- If the PI can justify risk/benefit to DURC/IBC committees and the U.S. government, should any GOF studies be prohibited?

SOURCE: Potter, 2016.

avoid creating a separate definition and policy oversight process for GOF research. Christopher Park from the Department of State stressed the importance of the scope of GOF studies of concern in covering both biosafety and biosecurity issues, which set it apart from the DURC process. Richard Frothingham expressed a concern that the DURC institutional process was considerably more cumbersome than the normal IBC process, and he would be reluctant to see them combined.

Dr. Frothingham highlighted the importance of the independence of review committees and their ability to access external expertise. He and Philip Potter discussed the advantages, for example, of including local public health officials in the membership of the IBCs, which both Duke and St. Jude do. Both also have regular contact with the Federal Bureau of Investigation's local Weapons of Mass Destruction coordinator. The importance of clear definitions was stressed by Diane DiEuliis from the National Defense University, while others, such as Mr. Park, highlighted cases where overly detailed definitions undermined the intended aim of the measure. There was a discussion about whether it was better to limit the scope of research likely to be captured under these definitions or, as proposed by Mr. Park, to have a fast-track process for removing research not deemed relevant during the review process. The unique nature of each research proposal was stressed by Dr. DiEuliis, as was the need to consider each proposal in context.

Several participants noted the importance of exploring alternative approaches to GOF studies of concern whenever possible, and the panelists discussed several specific examples of this happening. The value of broader expertise and nonspecialists in identifying alternative research approaches was noted in this regard by Drs. Frothingham and Potter.

Michelle Mello and several of the panelists felt that public trust was an important metric for assessing the efficacy of regimes for DURC and GOF studies of concern. Other participants suggested that reviews of DURC and GOF studies of concern were sensitive and should not be publicly available. Some participants argued that transparency was important and that relevant records should be made available. Both Dr. Frothingham and Gerald Epstein commented on the difficulties posed by the competing goals of protecting potentially sensitive information and ensuring transparency as part of gaining public trust. There were suggestions by Mr. Park, for example, that such information might be made available but not widely distributed.

The discussion also identified a number of tools that might strengthen future efforts. Gregory Koblentz of George Mason University highlighted the importance of learning from past experience. He and other participants called for mechanisms to capture lessons learned in a more systematic fashion. Professor Koblentz also called for additional help for PIs to understand the underlying concerns that drive assessments of DURC and GOF studies of concern. He also proposed more support to assist regulators in understanding what is possible at the laboratory level and to enable public understanding of the research.

There was also some discussion of the proposal from Silja Vöneky from the University of Freiburg to require laboratories to take out insurance against the risks of GOF studies of concern. Dr. Epstein saw the utility of this proposal—the Department of Homeland Security requires the laboratories it funds for Select Agent research to have insurance. He felt that it would be a useful approach for improving good behavior, although he saw challenges for insurance companies in developing accurate actuarial calculations on these risks.

The definition of GOF studies of concern contained in the NSABB WG's Draft Working Paper was revisited. The third criterion for defining GOF studies of concern was once again the most discussed by some participants, such as planning committee member Dr. Bloom from Harvard University, speaking in his personal capacity, finding no difference between the transmissibility criteria and the one connected to innate or acquired resistance to public health interventions. In connection to the third criterion, Dr. Bloom also raised issues of justice and equity around access to drugs in many parts of the world.

BEST PRACTICES TO INFORM POLICY DESIGN AND IMPLEMENTATION

Philip Dormitzer from the Pfizer Vaccine Research and Development Unit, a member of the Symposium Planning Committee, introduced the session as a continuation of the earlier plenary session on the U.S. policy landscape. This session would present the perspectives of several different key stakeholders, including regulatory agencies and the vaccine industry.

Michael Callahan from the Massachusetts General Hospital and Harvard Medical School opened by highlighting that U.S. efforts to balance the risks and benefits of GOF could be altered for use in other contexts and adapted to the needs of different countries. Dr. Callahan stressed the interconnected nature of the research and development enterprise in the life sciences and for biotechnology, asserting that "the world is flat for bio-innovation." As an example, Dr. Callahan noted that more viral pathogens have been sequenced in China in 4 months than have ever been sequenced in the United States and Europe.

A major theme of Dr. Callahan's remarks was that market-driven and beneficent GOF research is already happening around the world, all of which is outside of the U.S. and European policy and regulatory frameworks. One of his main points was that U.S. and European vaccine production does not always take cultural and other factors into account. For example, he noted that Western vaccines will only be used in Indonesia if they conform to requirements that make them halal. Another relevant example discussed was the production in Asia of effective and inexpensive H5N1 poultry vaccines. The life span of a chicken in Asia is about 6 months, so Western vaccines costing $7 per dose are not going to be used when Asian-produced vaccines costing pennies per dose are available. He also noted that countries in the group of 112 Non-Aligned Nations may refuse to share pathogen gene sequences with U.S. scientists because "they've been ripped off." He suggested that the U.S. government needs to protect "our international collaborators from R01-funded investigators who seek to do nothing more than get a virus, go home, and write their big *Nature* paper." Dr. Callahan concluded with a series of recommendations for aligning domestic and foreign policies relevant to GOF (see Box 3-4).

Robert Fisher from the Food and Drug Administration (FDA) introduced the FDA's main mission: to ensure that medical products and associated technologies are safe and effective. Dr. Fisher discussed a variety of regulations relevant to the evaluation of products and the implementation of regulatory mechanisms. He highlighted a number of approaches used by the FDA, including randomized clinical trials, surrogate endpoints, and animal efficacy data. He stressed that, regardless of the approach taken, the FDA relies on data for its decision making.

Dr. Fisher noted that the GOF framework focuses on specific agents of concern, or particular pathways of concern, potentially impacting

BOX 3-4
Recommendations for Aligning Domestic and Foreign Policies Relevant to Gain-of-Function Research

- Introducing Institutional Review Boards and Biological Weapons Convention guidance to foreign venture capitalists, incentivizing market entry through compliance.
- Licensing safe, effective, and inexpensive vaccines to foreign markets or exporting rational vaccine designs to foreign providers.
- Ensuring U.S. government–qualified expert review of academic claims for biosafety and pathogen research in foreign research facilities.
- Incentivizing host nation compliance through the use of metrics which demonstrate local benefit.

SOURCE: Callahan, 2016.

FDA-relevant research. However, the narrowing of focus to GOF studies of concern reduces the potential impact considerably. He provided examples of where GOF studies of concern might impact the work of the FDA, including the production of vaccine seeds from molecular clones, or adapting vaccine candidates to grow in cell-based systems rather than egg-based systems. He noted the potential for GOF concerns to impede rapid, large-scale production of vaccines to meet seasonal and emergency needs.

Lessons from the FDA's regulatory experience were also provided: for example, the importance of early and sustained engagement with stakeholders. Dr. Fisher stressed the value of ensuring sufficient flexibility in regulatory regimes and for their implementation.

Jonathan Moreno from the University of Pennsylvania framed his comments as both a bioethicist and a patient—or a "consumer" of the public health benefits of GOF research. Dr. Moreno identified a number of questions that he thought needed to be addressed when considering policy options for dealing with GOF research, including the impact of terminology and the response to the term GOF; the implications of mutants—both natural mutations and escape mutants; the potential for a generalized GOF policy being too broad to implement effectively; the adequacy of safety records for quantifying the risks of laboratory accidents; current levels of accident reporting; the need to address basic research and vaccine development activities where more acceptable or safer alternatives do not exist; determining the realities of the relationship between GOF methods and vaccine development; and the role of basic science during a health emergency.

Dr. Moreno identified five areas where he thought there was consensus relating to the controversy over research and the policy options to address it:

- Much regulation fails to hit the mark for this field and could needlessly delay vaccine development
- Some regulation is needed, for both biosafety and biosecurity
- Biocontainment does not have a perfect record
- Risk mitigation often only requires some imagination
- Sometimes there are acceptable alternatives to GOF studies of concern, even if they are not the best option

He also thought there might be agreement that GOF data alone cannot predict emergence of a pandemic (genotypes to phenotypes), but perhaps this is getting better; the long-term potential of pre-pandemic strain selection could be "transformative" in new vaccine development; humans are vulnerable to certain natural strains that could be targeted for research, such as bat SARS-like coronavirus strains; and animal model development for SARS and MERS should be permitted.

He highlighted opportunities for public deliberation as to whether all three of the characteristics for GOF studies of concern, as proposed by the NSABB, are needed for an experiment to warrant additional oversight, or whether the production of a pathogen anticipated to possess two of the characteristics would be sufficient. Dr. Moreno highlighted the need to build on best practices when developing capacities to review GOF studies of concern. He reviewed the composition and mission of the Wisconsin Bioterrorism Task Force and the Stanford University benchside ethics consultations as examples to be considered. Dr. Moreno presented a potential model for institutional bodies for the operational review of GOF studies of concern: Risk–Benefit Assessment Teams, or R-BATs (see Box 3-5).

Ethan Settembre from Seqirus provided an overview of the global influenza system that addresses variability in influenza viruses to develop and deliver candidate vaccine viruses. Dr. Settembre described an example of how the system works in practice, detailing vaccine generation over a 4-5 month period in response to the H1N1 influenza pandemic in 2009. He noted that while vaccines can be produced increasingly quickly, pandemics

BOX 3-5
A Proposal for Risk–Benefit Assessment Teams (R-BATs)

Risk–Benefit Assessment Teams should work in an informative and consultative (but not dispositive) manner. The team should encourage researchers to demonstrate that they have considered the risks and benefits at the current stage of their work. These teams are intended to move beyond a paper mechanism to a dynamic, real-time process. In particular, they should

- Be independent and multi-disciplinary.
- Represent the perspectives of both science and security communities.
- Work through an iterative process that spans the life cycle of the research.
- Use a schedule based on milestones and perhaps be able to make unannounced audits.
- Make assessments based on case-specific risk–benefit parameters.
- Help to develop and disseminate best practices for research with GOF studies of concern.

Further consideration would be necessary to determine whether their existence should be voluntary or mandatory.

SOURCE: Moreno, 2016.

emerge even more rapidly, necessitating further research and development. Dr. Settembre highlighted a need to further speed up production of vaccines in response to both pandemic and seasonal influenza events. He discussed a synthetic process for generating vaccine candidates using attenuated backbones and available hemagglutinin and neuraminidase sequences that was developed with the J. Craig Venter Institute and Synthetic Genomics Vaccines, Inc.. The process allows the generation of synthetic influenza viruses that are attenuated, but would allow for speed, accuracy, and high yield. He noted that this is one of the ways to make vaccine viruses to address immediate important medical needs in a short period of time to get ahead of the wave of infection. This approach had been used to produce an H7N9 vaccine candidate in 2013.

Discussion

The discussion that followed included both an interchange among the panellists and questions and comments from the audience. Among other topics, the discussion returned to considering different ways of defining GOF studies of concern. Philip Dormitzer pointed out that not all GOF research involves GOF studies of concern, and therefore not all the research needs to be overseen by any additional policy frameworks. Following his comments, panelists and participants discussed the subjective nature of determining what is (and what is not) of concern. Michael Callahan argued that any definition for GOF studies of concern needs to be general, adaptive, and culturally appropriate for foreign scientific communities. Participants noted that specialist terminology might not translate well into other languages and settings: for example, Dr. Callahan noted long-standing issues around the meaning of "biosecurity."

Participants also reexamined the importance of ensuring that the process to consider policy options, as well as any new frameworks it produces, cover both the public and the private sector. The importance of adequate containment for GOF studies of concern was another reoccurring theme during the discussions. The issue of enforcement was also revisited. Possible unintended consequences for new policy frameworks to oversee GOF studies of concern were discussed by Dr. Callahan, and possibilities of a negative impact on vaccine production were considered. The risks of regulatory uncertainty were also addressed, with some participants arguing that regulatory burdens are more acceptable when the "what" and the "why" are clear. Issues around the harmonization of domestic oversight regimes, such as those for DURC and GOF studies of concern, were also highlighted by Robert Fisher. Some participants called for the development of a more overarching framework to deal with risks and benefits from life sciences research.

The international implications of determining thresholds of concern or acceptable risk were considered, as well as international perceptions about why the United States might be concerned about this research. Dr. Callahan suggested that it was important to understand the nature and motivation of relevant international stakeholders to improve the dialogue on GOF research. To this end, he noted the importance of strengthening research collaborations, in particular working more closely with partners inside their countries. Gregory Koblentz asked whether it was time to move beyond stovepiped concepts of "biosafety" and "biosecurity" to adopt a more holistic concept of "biorisk management." Dr. Fisher responded that, from a regulator's viewpoint, to the extent that such an approach could reduce uncertainty, it could be helpful.

There was an exploration of the impact of over-regulating GOF research for countermeasure development. Dr. Dormitzer pointed out that, because one of the factors for identifying GOF studies of concern is the absence of effective countermeasures, limiting research that could provide such measures could be counterproductive. There was also consideration of the opportunity costs of not doing research, especially in justifying potential barriers to developing countermeasures. Issues around intellectual property were also explored, with Dr. Callahan discussing barriers for the development of countermeasures, or barriers to the conduct of science internationally. There was a call to change the incentives for countermeasure development—to produce more players, more stakeholders, and therefore more solutions. To this end, Dr. Callahan recommended that greater attention be paid to foreign industry, as an increasing number of products and self-sufficient markets were being developed.

Challenges in disease surveillance were discussed. While some participants suggested that knowledge produced by GOF research could be useful for detecting emerging pathogens, others noted the lack of current surveillance capacity. Current shortcomings in data sharing and capacity for disease surveillance can also distort risks from disease and the impact of public health measures, according to Dr. Callahan. He also noted international concerns that disease surveillance data and capabilities are being used for nonpublic health purposes.

4

International Policy

INTERNATIONAL DIMENSIONS OF GAIN-OF-FUNCTION RESEARCH

Barry Bloom from Harvard University, a member of the Symposium Planning Committee, introduced the session by noting that, as had already become clear in earlier discussions, the issues related to gain-of-function (GOF) research were not confined to the United States. This session would provide background and insights about the international dimensions of GOF research and illustrate the ways in which various organizations outside the United States have been contributing to the discussions from the beginning.

Ruxandra Draghia-Akli from the Health Research Directorate of the European Commission introduced the European Union (EU) innovation framework Horizon 2020. With a budget of €79 billion, the program is intended to support research and development that is increasingly complex, interdisciplinary, and costly, and that also requires a critical mass. It provides a vehicle for increased collaboration across the 28 countries of Europe. The Horizon 2020 framework covers a broad range of research and development activities; most relevant to GOF research is the section on Societal Challenge 1: Health, Demographic Change and Wellbeing, with a budget of €7.4 billion. The research funded under this framework has to have civil or public health applications. Any dual use potential is unintentional.

There has been no specific call for proposals on GOF research according to Dr. Draghia-Akli, but under the health calls, proposals were encouraged to strengthen research on prediction, identification, modeling, and surveillance of newly emerging infectious diseases in humans, and to identify factors promoting the emergence of pathogens with human pandemic potential from pathogens with a zoonotic background and related prevention strategies. Both of these areas could potentially result in proposals involving GOF research. Five EU-funded research projects with GOF elements were identified: (i) EMPERIE (European Management Platform for Emerging and Re-emerging Infectious Disease Entities, 2009-2014); (ii) PREDEMICS (Preparedness, Prediction and Prevention of Emerging Zoonotic Viruses with Pandemic Potential using Multidisciplinary Approaches, 2011-2016); (iii) ANTIGONE (Anticipating the Global Onset of Novel Epidemics, 2011-2016); (iv) AntiBotABE (Neutralizing antibodies against botulinum toxins A,B,E, 2010-2015); and (v) TIRAMISU (Humanitarian Demining Toolbox, 2012-2016).

Dr. Draghia-Akli then outlined the ethics review processes that research undergoes in Horizon 2020.

- During proposal preparation, applicants are asked if their proposal has an exclusive civilian focus on research and if their research uses or produces goods or information that will require export licenses in accordance with legislation on dual use items.
- Ethical screening is carried out for each successful proposal by at least two ethics experts, drawing on a variety of different backgrounds, including law, philosophy, medicine, and biology.
- A full ethical assessment for all proposals containing potential dual use issues is carried out by at least five ethical experts.
- At the end of the whole process, the ethics report determines if the project has clearance, requiring no further action; conditional clearance, requiring changes to be made to the description of work (such as requirements for permits, follow-up, or ethical audits); or no clearance, meaning that the project will not be funded.

There has also been specific dialogue with stakeholders in the European Union on GOF research, including the European Society for Virology, which has a common policy for scientific research and publications; the Foundation for Vaccine Research, which has called for a comprehensive risk–benefit assessment of GOF studies of concern; and the European Academies Science Advisory Council (EASAC), which established a working group in autumn 2014 to explore consensus on key questions, identify further GOF issues, and clarify options for policy development. The European Union acknowledged the need to improve awareness and best practices among members of the scientific community and to promote an underlying culture of responsibility, given the potential for accidental release and misuse. The European Union also welcomed the EASAC working group recommendations (EASAC, 2015a; see the comments by Volker ter Meulen).

Dr. Draghia-Akli provided three different approaches toward implementation within the European Union: researcher-based approaches, as used in the United Kingdom; researchers being overseen by institutions, as used in the Netherlands; and supervision by a national agency, as used in France.

Volker ter Meulen, the chair of the EASAC working group, introduced his institution. EASAC was formed in 2001 to enable European national academies of science to collaborate in giving advice to EU policy makers (e.g., the European Commission and Parliament). Its membership comprises all EU national academies of science plus Norway and Switzerland, and its objective is to deliver consensus outputs to provide a means for the collective voice of European science to be heard. Dr. ter Meulen explained that the EU scientific community had expressed differing views to the president of the European Commission in 2013 on relative benefits and risks of GOF influenza virus (H5N1) research (European Society for Virology, 2013; Foundation for Vaccine Research, 2013). As a result, the European Commission and its chief scientific adviser requested EASAC to clarify and advise on these issues. EASAC brought together scientists, nominated by its member academies, who represented a mix of expertise and a wide range of views about the GOF controversy. The group sought to find areas of consensus as well as issues that had not been resolved. Its report also offered recommendations about what further analysis would be necessary to assess future options for research with potentially pandemic pathogens. The report also identified which of the European Union's current regulations applied to GOF research, how national- and EU-level responsibility should be divided, and what best practices already exist at the national level that could inform other countries.

During the course of this work, EASAC identified a range of critical issues to consider when addressing GOF research as well as key messages, which are summarized in Box 4-1.

EASAC has subsequently produced messages to academies of science worldwide, policy makers in EU institutions, and EU member states as well as to research funding bodies, regulatory bodies, professional societies, and others in the scientific community. It has also worked to catalyze further broad engagement via member academies.

Dr. ter Meulen concluded by providing some insights for strengthening international consideration of GOF issues, including the importance of addressing differences in understanding and in systems between countries and regions; ensuring layered, integrated approaches; building

BOX 4-1
Key Messages from the European Academies Science Advisory Council's Work on Gain-of-Function Research

Self-regulation and harmonization
- Good practice requires conforming with regulations, safety provisions, codes of conduct, and justifying proposed research.
- Self-regulation means instigating a series of checks and balances on research within the scientific community. It requires raising awareness among researchers and their institutions, thereby necessitating education.
- Attention to biosafety issues is needed at all stages of the research life cycle.
- There is a continuing role for Academies of Science in promoting biosafety and biosecurity norms and supporting audits of research practices.

Benefit–risk assessment
- This is not a "once and for all" calculation but a continuing, collective commitment to understand and communicate the issues.
- Incommensurable parameters measured in risk and benefit do not allow a value-free determination to be made.
- Questions remain as to the feasibility of quantifying benefit as prospective public health impact or describing its impact on the generation of scientific knowledge.
- Academies and learned societies need to continue to promote discussion across the scientific community and with other stakeholders.

EU/national activities and organizations in biosafety and biosecurity
- There is a possible role for the European Commission (DG Sante) Health Security Committee in collating available information.
- Guidance is needed for research funded by Horizon 2020 as well as at the national level.
- All researchers and institutions need to conform with EU regulations as implemented nationally.
- No new EU-level body was recommended.
- Member states should have clear national advisory approaches and governance mechanisms with statutory powers.

Publication of sensitive information
- Researchers and their institutions all have responsibility to make decisions about publishing sensitive information.
- Journals should be encouraged to seek appropriate advice, including from security experts.
- Export control regulations are an inappropriate and ineffective vehicle to block publication.
- The European Commission's (DG Research) attempts to raise awareness about revision of these regulations are welcome—researchers should continue to inform policy-makers about these issues.

continued

BOX 4-1 Continued

Public engagement

- Trust and openness are crucial for researchers and their institutions.
- Academies and others in the scientific community should actively participate in public dialogue—articulating objectives for research, the potential for benefit and risk, and biorisk management practices adopted.
- EASAC is committed to continuing working with academies to promote engagement.

SOURCES: EASAC, 2015a,b; ter Meulen, 2016.

links between researchers, policy makers, and other stakeholders; and a continuing commitment to public engagement.

Silja Vöneky from the University of Freiburg discussed the German Ethics Council (GEC) report on biosecurity from 2014 (German Ethics Council, 2014). Dr. Vöneky introduced the GEC, noting that it is an interdisciplinary independent counsel of experts whose 26 members are appointed by the president of the German parliament. She reviewed recent work exploring options for biosecurity for research and health and noted that the GEC report explored biosecurity issues but not biosafety because the regulatory regimes associated with biosafety are much more developed.

Dr. Vöneky presented five recommendations for future GOF research based on the findings of the GEC report (see Box 4-2). These recommendations reflected those proposed for use within Europe, focused on five different areas of governance: raising the level of awareness on biosecurity among the scientific community; elaborating national biosecurity codes of conduct; reviewing research funding; making specific national recommendations tailored to national needs; and developing European and international initiatives.

In the German context, Dr. Vöneky recalled that the GEC had recommended that an appropriate definition of the research of concern should be included in an act of parliament; the definition should be further developed by detailing relevant groups of experiments in a statutory instrument or regulation; and a list of agents associated with the research should be developed. She commented that the list of agents will need to be updated to reflect advances in the life sciences, suggesting that it should not be listed in legislation.

Dr. Vöneky noted that it was difficult to assess the impact of these recommendations over the year since they were made. She highlighted

BOX 4-2
Recommendations for Future GOF Research Based on the Findings of the GEC Report on Biosecurity

Raising awareness in the scientific community—To promote responsible research and improve knowledge of, and access to, relevant resources. One approach to raising awareness was to integrate biosecurity components into undergraduate and graduate life science curriculum.

The use of codes of conduct—Codes were deemed to be practical tools to define responsible approaches for dealing with biosecurity challenges, including by detailing concrete obligations to minimize risk. They were felt to be useful instruments for self-regulation and can be supplemented by broader standards. Codes of conduct could be adapted to address GOF issues, including thresholds for GOF studies of concern.

Strengthening the role of research funding in ensuring responsible conduct—Funding of GOF research should require adoption and adherence to the above code of conduct. Specific funding guidance should be developed ensuring that GOF studies of concern are not funded when there is no need to use GOF approaches or when the risks outweigh the benefits.

Establish a new commission to oversee GOF studies of concern – An independent body, with interdisciplinary membership and participation by civil society, should define GOF studies of concern, conduct risk–benefit analysis of specific research proposals, decide on any additional measures to mitigate or manage risks associated with the research, and undertake relevant consultative roles. It should become a legal obligation to consult the commission before undertaking GOF studies of concern.

Regional and international engagement—Common standards can play an important role in addressing biosafety and biosecurity concerns related to GOF studies of concern. Efforts within scientific communities should continue to develop a common understanding on what constitutes responsible research. An attempt should be made to develop an international code of conduct. At a regional level, States should advocate for a common position on the funding of GOF studies of concern. States should also work internationally to define and classify dual use research of concern (DURC) and GOF studies of concern and appropriate biosafety and biosecurity precautions for undertaking such work. A new international instrument to define the fundamental principles and limitations of GOF studies of concern should be negotiated. This could be a formal treaty or more likely a soft law instrument.

SOURCE: Vöneky, 2016.

progress in Germany in the promulgation of codes of conduct to address DURC issues, and she discussed the implementation of such codes in a number of research institutions.

Dr. Vöneky also highlighted a number of other results that might be relevant to GOF discussions. She believed that soft measures, such as requirements connected to funding, might be less suitable in the EU context. She suggested that measures to evaluate and manage risks associated with GOF research would need to be either codified by states into appropriate laws and regulations or contained with other legal frameworks such as constitutions or international treaties because of competing interests between the rights and freedoms of science and scientists and rights associated with the right to life and health for other parts of the population. Dr. Vöneky noted that the work undertaken by the GEC revealed that existing legal rules governing GOF research in Germany and Europe are insufficient and need to be more coherent. She highlighted internal inconsistencies around the publication of results and funding arrangements.

Dr. Vöneky concluded by stressing the need to balance scientific freedoms and responsible research with the need for proportional measures that do not unnecessarily impede research but do help to manage risks. She reflected on thresholds for GOF studies of concern, suggesting that such a concept might usefully capture experiments that might result in pathogens, which increase the danger of an epidemic of a severe human disease. She felt that such experiments should not be undertaken unless a direct, concrete, and overwhelming benefit for life or human health is probable.

Keiji Fukuda from the World Health Organization (WHO) explained that the world is currently facing a broad mix of issues and uncertainties related to genetic technologies and their potential to do harm, such as GOF research. Other approaches, such as synthetic biology, offer ways to generate novel organisms. Furthermore, Dr. Fukuda noted that the nature of these challenges is evolving as access to the necessary technologies changes: for example, through the emergence of cheaper technology and the advent of private community laboratories. Dr. Fukuda also noted that the entry into force of the Nagoya Protocol on sharing the benefits of biological resources also impacts this space. Furthermore, he recalled that developments, such as the WHO Pandemic Influenza Preparedness Framework deal with not only the movement of viruses but also the movement of sequence information. Dr. Fukuda noted that while these are critical international issues, global awareness of them and how they intersect remains minimal. He highlighted the lack of a clear strategy and outstanding questions as to whether they should they be dealt with separately or were better addressed together. Dr. Fukuda recalled that while risk assessment can be a scientific and precise process, risk perception, tolerance and management are cultural, political, and, at the global level, consensus based. Dr. Fukuda presented four options for further work on GOF research (see Box 4-3).

Discussion

Barry Bloom led a moderated discussion among the panelists. They explored opportunities for interaction between U.S. and European efforts to address GOF research, with Dr. Draghia-Akli expressing how beneficial such exchanges on common policy problems could be. The importance of information exchange was repeatedly expressed. One panelist felt that the European Union was likely to be flexible about approaches to the biosecurity aspects of GOF research, but noted that given the highly developed arrangements already in place there may be less opportunity to influence biosafety policy. The complexity of the European regulatory architecture was also noted, with one panelist suggesting that additional measures were added but rarely replaced existing arrangements. A trend toward the European Union engaging international partners was highlighted, especially through the development of principle-based voluntary frameworks that could be implemented by partners. Past collaborations between the United States and the European Union were noted on health and biomedical related policy development: for example, bringing together funding agencies to streamline work on rare diseases. Past examples also included collaboration on sensitive issues, such as data and sample sharing.

BOX 4-3
Four Options for Further Work on GOF Research

- Options for going forward will necessitate developing global consensus on technical aspects, such as issues, principles, definitions, and terminology.
- At an operational level, there are examples of programmatic activities implemented by WHO that might offer models, such as the prequalification of laboratories, the oversight of smallpox research and the inspection of the laboratories conducting research.
- Multilateral forums such as the Biological and Toxin Weapons Convention or the International Health Regulations might be suitable venues for addressing specific aspects of GOF but will likely be time consuming.
- Member state funding and support is essential regardless of what approach is taken.

SOURCE: Fukuda, 2016.

Panelists also considered options for attempting to ensure that GOF studies of concern were conducted only under appropriate safety conditions. At the suggestion of Barry Bloom, panel members discussed precedents used elsewhere for the prequalification of appropriate laboratories, assessing them against predetermined capabilities: for example, those used for quality control of laboratories used by United Nations agencies. Several participants supported such an approach but highlighted that it would be necessary to consider carefully what the desirable capabilities would be. Other participants felt that the GOF studies of concern context was considerably more complicated than the purposes for which prequalification has been used in the past, and they suggested that the desirable capabilities would be too context dependent for such an approach. They also noted that the number of relevant facilities might be larger than those found in other areas where prequalification has been used.

An open discussion followed and consideration of prequalification of laboratories continued. Gavin Huntley-Fenner questioned which international organization might oversee such an approach. WHO, the International Standards Organization, and the United Nations Educational, Scientific and Cultural Organization were discussed. Keiji Fukuda stressed the importance of any hosting organization being perceived to be neutral and having the trust of key stakeholders.

Issues around standards and harmonized approaches were also explored. Harvey Fineberg, chair of the Symposium Planning Committee, noted that in certain cases—such as for the approval of medicines, drugs, and other medical devices—there was still a notable degree of difference in what is approved, and when, despite a comparatively common agreement on the characteristics to be assessed, relatively straightforward measurements, and well-established decision-making processes. Other participants noted that in the European context, while risk assessment might be carried out collectively, regulatory approval still happened at the national level. Some participants suggested that the chances of creating a common system for GOF studies of concern in the short term were small, especially given the absence of a common definition. Other participants noted that the number of scientists and laboratories potentially conducting GOF studies of concern was currently limited and that there might be opportunities to develop common standards— for example, for biosafety precautions—among the relevant community.

The possibility for developing common approaches between the United States and Europe was also explored, with Silja Vöneky suggesting that reaching such an agreement might help jump-start a broader international process. Michael Callahan felt that a broader buy-in from the start would help legitimize the process. To underscore that argument, Piers Millett from Biosecure transmitted the views on GOF of the 112 states that comprise the Group of Non-Aligned Movement and Other States under the Biological Weapons Convention (BWC) by reading aloud from the Group's statement to the BWC.[1]

There was also an exploration of whether harmonization efforts should be scientist-led or state-led, with different participants favoring different models. Some participants noted that, at present, GOF studies of concern were largely confined to public institutions, enabling governments to play a leading role. Others noted that only a limited number of states had so far shown an interest in GOF studies of concern, suggesting that a scientist-based, bottom-up approach may help increase government interest around the world. There were also discussions of whether a formal approach was needed, requiring international instruments, or a more informal approach might be more suitable, perhaps through appropriate guidelines such as those used to underpin international efforts on infection prevention and control. Participants also discussed the value of strengthening a culture of responsible research among relevant scientific communities, noting that they had key insights into the risks associated with GOF studies of concern.

The changing distribution of research capacities, sources of funding, and the markets they serve were also discussed. Some participants noted that these developments complicated efforts to address GOF studies of concern, and others noted that it required additional efforts to understand a broader variety of motivations for and contexts within which GOF research might be conducted. Participants discussed incentivizing industry participation, alternative funding strategies and business models in general, and public–private partnerships in particular, for dealing with changes in markets, funding and demographics. Participants provided a number of examples of successful precedents, including the Innovative

---

[1] The statement is "there have been recent advances demonstrating the increasing sophistication of synthetic biology, together with other enabling technologies, which have benefits, together with the potential for uses contrary to the provisions of the Convention. All states must conduct such activities in a transparent manner, in order to build the confidence of other States Parties. There is a need to regulate these activities, to ensure that they do not lead to any concerns related to ethics, safety and security as well as any uses contrary to the Convention. This has assumed added importance in the light of reports concerning experiments that have been taking place on highly contagious virulent flu strains like H5N1, as well as the production of several new strains of viruses that are both contagious and deadlier than the 1918 Spanish flu that killed almost 50 million people, and the discovery of the deadly smallpox variola virus dating back to the 1950s. Such regulation must, however, be undertaken in a manner that does not hamper scientific and technological developments that are in keeping with the spirit and letter of the Convention, which are of benefit, more especially to developing countries." It is available at http://www.unog.ch/80256EDD006B8954/(httpAssets)/DF2D9E3CAA6D5FEDC1257EA400369E6E/$file/NAM+Statement+on+S&T+MX+2015-3+final.pdf.

Medicine Initiative in the European Union, and global networks for building preparedness for emerging epidemics.

OPPORTUNITIES TO HARMONIZE GOF RESEARCH POLICY AND PRACTICE

Ronald Atlas from the University of Louisville, a member of the Symposium Planning Committee, introduced the session. The plenary on the first day provided participants with an awareness of the international context within which the GOF controversy has evolved. The purpose of this session was to look ahead, to explore the potential for increasing international coordination of policy and practice for GOF studies of concern. What are the opportunities in different regions, including those where the research is performed and those where the pathogens of concern are endemic? What roles might national governments take in fostering efforts at coordination? What are some of the international venues, such as regional or international organizations, where discussions could take place and policy options could be developed? What are the roles for national and international scientific organizations?

George Gao from the Chinese Academy of Sciences and the Chinese Center for Disease Control and Prevention began by discussing risk and benefit. Dr. Gao discussed an example of the H7N9 influenza virus, noting that study of the virus could be directed at finding a mutation in the receptor binding site that might be responsible for allowing the virus to switch from an avian to a human host and allow the virus to transmit to humans. Dr. Gao noted, however, that checking all possible genetic combinations is infeasible, and, given the importance placed on finding the mutations responsible, a GOF approach proved most efficient.

Dr. Gao stressed the importance of international collaboration, cooperation, and harmonization. He recalled one case where two researchers, one in the United States and another in China, were collaborating on research connected to Golden Rice. The contents of the underpinning agreement were different in Chinese and English, which led to misunderstandings and substantial impediments to the research. He felt that harmonization was needed on more than just policy development and that it was necessary to have oversight of the research being undertaken. He suggested that it is important to monitor what is happening in laboratories. Dr. Gao noted the need for a suitable international forum for discussions at the government level. He felt that it was important for top officials in many countries to engage with this issue. He suggested, however, that in many countries there were still opportunities for greater domestic harmonization of relevant rules and regulatory approaches.

The interests of scientists often drive the direction and approaches to research, Dr. Gao noted. Therefore, he felt that it was important to engage with individual researchers on these issues. He believed that GOF experiments should be done in highly regulated laboratories and undertaken only by the best scientists.

Gabriel Leung from The University of Hong Kong began by considering GOF research in context. He noted that this was a discussion of risk to humans and, to a lesser extent, ecological security. He underscored that this was an international issue because pathogens do not respect borders. Dr. Leung suggested this is also a global security issue, not only a U.S. national health security concern. He felt that the primary outcome of policy discussions should be conclusions as how best to keep the global population safe from potential consequences of pathogens that were highly virulent, highly transmissible, and/or resistant to public health interventions.

Hazard analysis, according to Dr. Leung, was a critical control point. He highlighted lessons that might be learned from food safety experiences. Dr. Leung suggested that it was necessary to look for the weakest link in the global supply chain and argued that, in the case of GOF studies of concern, it was a lack of public health preparedness. He recalled that the majority of countries around the world had self-declared their inability to meet core requirements under the International Health Regulations. He suggested that investing in global capacity to respond to disease minimizes the proportion of GOF studies of concern that would then be of concern. To this end he commended the recommendations of the recently released report of an international commission hosted by the National Academy of Medicine on which he had served (Commission on a Global Health Risk Framework for the Future, 2015).

Dr. Leung noted that a highly organized regime for governance of GOF research was important for obtaining human security. He felt that the issue had been largely ignored elsewhere in the world. While he acknowledged that a national policy in the United States would have a global impact, he noted that its relevance should not be overestimated. Stringent arrangements in the United States, or a ban on GOF studies of concern, according to Dr. Leung, would not stop risks to the global population from research carried out in other countries.

Dr. Leung also argued that overly burdensome regulations can lead to unanticipated consequences—perhaps driving GOF research underground or possibly relocating it to other countries without such regulations. He expressed concern over how the broad findings of the National Science Advisory Board for Biosecurity might be translated into guidance and implemented by IBCs in institutions. He felt that greater clarity, especially as to what is (and what would not be) permitted, was needed, as was more guidance on the implementation of the proposed policy framework.

Dr. Leung provided a number of specific reactions to the inputs to the symposium (see Box 4-4) and concluded that responsible science with robust oversight of GOF studies of concern is warranted, but it should "not squeeze the lifeblood out of scientific enterprise." He felt that the balance between the two must be clearly defined and continually fine-tuned.

Nisreen AL-Hmoud from the Royal Scientific Society of Jordan noted that understanding life processes is becoming ever more important in terms of health, nutrition, and industrial application. Dr. AL-Hmoud suggested that the Middle East and North Africa (MENA) region lags behind other parts of the world in addressing issues in life sciences research.

BOX 4-4
Issues for Further Consideration When Considering Opportunities for International Harmonization of Approaches to GOF Research

- Is there any unique value or value-added from GOF studies of concern or can alternative methods exhaustively derive the same knowledge set?
- GOF research has led to new characterization of pathogens as well as follow-up research identifying new markers of mammalian adaptation.
- Is the proposed third criterion of a GOF study of concern truly orthogonal to the other two dimensions of transmissibility and virulence? Is there an advantage, for example, in better defining or making more encompassing the two truly orthogonal axes of transmissibility and virulence as the product of host-agent interactions as opposed to just innate properties of the agent alone?
- Do we need another layer of regulations or could existing regimes be adapted?
- More consideration may be warranted as to the unintended or intended consequences of policy options. Particular focus should be placed on avoiding such heavy burdens on GOF studies of concern that the research is avoided altogether.
- Financing has always been a powerful modifier of behavior and offers opportunities for shaping engagement on GOF studies of concern.
- Conflicts of interest should be avoided. It is to be hoped that situating the bodies reviewing research for GOF studies of concern inside the same agency that funds the research does not provide a conflict of interest.
- The RBA (risk and benefit assessment) is far from a trivial exercise but more work is necessary to definitively resolve the original questions posed. While it is difficult to draw direct lessons from an RBA exercise, we are now in a better place to understand what we do not know and how much we need to learn.

SOURCE: Leung, 2016.

She stressed the importance for the region of greater progress in ensuring that natural diseases are contained as soon as possible; that harmful consequences of research are minimized; and that laboratories operate safely—both for their workforces and for the communities in which they are situated.

While controversies around research involving highly pathogenic avian influenza virus and the Middle East respiratory syndrome coronavirus (MERS-CoV) have generated considerable discussion and debate among virologists, public health scientists, and experts in the United States and certain other parts of the world, a considerable need for raising awareness about GOF research persists in the MENA. This is needed for laboratory directors and policy makers as well as for life scientists. For maximum benefit, Dr. AL-Hmoud argued that policies and practices aimed at reducing and managing risks should be planned in a holistic manner as part of national safety and security strategies. She noted that, while some countries have begun to develop such plans, many others have not.

Dr. AL-Hmoud noted that while risks vary from region to region, and from one country to another, without a common methodology for assessing risks and for having appropriate policies and practices to manage and mitigate these risks, any international effort will be neither comprehensive nor effective. The countries in the region have explicitly recognized the need for comprehensive scientific strategies, and Dr. AL-Hmoud reviewed efforts under the Biosafety and Biosecurity International Conference series as an example.

Dr. AL-Hmoud stressed the public health impact of coronaviruses, in particular MERS-CoV, and she recalled that the antigenic relationships among the different coronaviruses or how these relationships influence the capacity of different strains to emerge in human populations remains uncharacterized. While she noted progress in relevant tools and information, Dr. AL-Hmoud also stressed that important research questions need to be explored further and they must address potential issues around security and select agent status. She highlighted opportunities for the MENA region to learn from the experiences of other regions, to adopt best practices, and to develop networks of experts.

Dr. AL-Hmoud discussed the importance of developing systematic programs that strengthen human capacity for safe and secure handling, importing, and exporting pathogens to strengthen the oversight of GOF research. She highlighted the need for certain infrastructure and policies at the national level. Dr. AL-Hmoud suggested that these programs should offer considerable regional and international benefits by reducing risks from pandemics and epidemics, regardless of whether they are natural, accidental, or deliberate. To reduce the risk of biological accidents, Dr. AL-Hmoud called for better safety standards and practices, and improved designs and procedures for security systems at biological facilities. In relation to GOF research, Dr. AL-Hmoud also noted the need for better education and training, more awareness raising, detailed consideration of unintended consequences, and broader adoption of best practices and codes of ethics.

Michael Selgelid from Monash University in Australia suggested that too much of the deliberative process and decision making on GOF research had been restricted to scientists. He argued that there had not been sufficient involvement of the general public and that there was a need for greater engagement of a wider range of stakeholders, including those from other countries.

Dr. Selgelid felt that some policy decisions, especially on risks affecting the global community, could only be made by an international body. He suggested that GOF research poses issues of global justice, including sharing of the benefits of this research. If the risks are universal, there may be issues if the benefits are available only to some of the countries. Dr. Selgelid felt this was particularly important for medical countermeasures. He also noted that a decision to conduct GOF studies of concern in only maximum-containment facilities would effectively preclude the majority of countries from undertaking such work. Dr. Selgelid suggested that WHO was the most legitimate international body to make decisions about GOF research. He discussed the possibility of creating a new WHO committee, similar to the body that oversees smallpox research, to undertake such a task. Dr. Selgelid also discussed the possibility of developing a new stand-alone body for the oversight of GOF studies of concern.

Dr. Selgelid noted that some countries are more likely to be exposed to risk from GOF research than others, especially where vaccines or therapeutics available in richer countries are not available in poorer countries. He noted that differences in access to basic healthcare could also result in an uneven risk distribution from GOF studies of concern.

He highlighted a number of lessons from earlier discussions of DURC. For example, he reviewed findings from the 2010 WHO guidance document *Responsible Life Sciences Research for Global Health Security* (WHO, 2010) (see Box 4-5). Dr. Selgelid also identified a number of approaches for harmonizing GOF research policy. He noted possibilities for gathering greater input from other countries bilaterally. He discussed a collective international harmonization process to create a level playing field. He also considered a more formal international governance regime for policy making and decision making. Dr. Selgelid reviewed the possibilities for using different frameworks, either by strengthening existing treaties or by creating new international agreements or compacts. He cautioned that these more formal arrangements would be difficult to achieve and involve a great deal of work. He also noted more standards-based approaches to

BOX 4-5
Insights for GOF Policy Making Drawn from Past Discussions on Responsible Life Sciences Research for Global Health Security

- There will not be one-size-fits-all approaches to managing risks—some solutions make more sense in some places than others.
- Opportunities to leverage existing regulations and governance structures should be explored whenever possible.
- If additional measures are needed to deal with GOF, expanding existing committees overseeing research might be considered rather than building new ones.
- There are different levels that can be used to address concerns over research: some approaches can be undertaken by individual scientists; other activities can be conducted at the institutional level; professional scientific bodies need to make certain decisions (such as on codes of conduct); other decisions need to be made by domestic governance (such as on regulations or education); and funding bodies can play an important role by taking risks into account when making funding decisions.

SOURCE: Selgelid, 2016.

governance, discussing the framework in place governing human subjects research and suggesting that the ethics governance regime might be expanded to include oversight of GOF studies of concern.

Discussion

The discussion that followed expanded on ideas and concepts introduced during the presentations. Participants explored the comparative advantages of using a standards-based approach based on the existing ethics governance regime. Michael Selgelid again argued this might be easier than a treaty-based approach because it could take advantage of existing policy frameworks and offer logistical benefits. Participants also discussed whether international harmonization might be best achieved through international organizations or international scientific bodies. Some participants felt that both approaches should be pursued in concert. Opportunities for using insurance requirements to harmonize GOF approaches were also discussed, and David Stanley introduced a concrete proposal from the Future of Humanity Institute to utilize the grant making process to address potential risks (Cotton-Barratt et al., 2016). The institute proposed to "price the expected value of any damages that could result from GOF research into the price of the grant being considered.

Then they could either require grantees to purchase liability insurance to cover the possible damages from this or, alternatively, require a payment to the state or non-state body to cover the expected cost of that research."

The reasons for seeking international input and harmonization were explored. Christopher Park from the Department of State outlined three objectives for seeking greater interaction, including (i) to get greater clarity as to foreign views on U.S. measures; (ii) to change behavior of individual researchers, perhaps best achieved through international scientific bodies; and (iii) trying to change behavior of other governments, requiring different approaches either through multilateral settings or coalition building. He also noted that if the intent was to address laboratory biosafety issues, it would require engaging one set of actors in associated settings, while a separate community and associated forums would be necessary for addressing biosecurity information risks. Another participant highlighted the importance of engaging the human and animal health communities, given the zoonotic nature of relevant diseases.

Keiji Fukuda from WHO stressed the importance in successful international efforts of a common understanding of the nature of the risk being addressed. He offered the negotiation of the International Health Regulations, the WHO Pandemic Influenza Preparedness Framework, and measures to address antimicrobial resistance as examples. He then suggested that such an international common understanding does not exist with regard to GOF studies of concern and that international engagement might be better focused on reaching a technical agreement on the nature of the risk posed by this research.

Participants also discussed three options for balancing national action against a broader international approach: to act now solely at a national level; to act now at a national level but send a clear message as to the desirability of subsequent international engagement; or to begin working on a full international policy process from the outset. Several participants felt that, given the international nature of the risks being addressed, the first option was not appropriate for GOF research. The same participants suggested that the decision as to whether the second or third approach was more suitable should be based on the resources available, the preexisting levels of international concern, and the level of need for international consensus. Another participant suggested that if sufficient resources could not be secured from the outset, it might be better not to initiate an international process rather than have to abandon it after a short while. Some participants highlighted the value of the Global Health Security Agenda as a model for building an international partnership with opportunities to shape the process.

One question raised during the discussion was who should determine the criteria for classifying research as GOF studies of concern or for identifying specific research proposals that meet those characteristics. Dr. Selgelid suggested that it might usefully be based upon a mutlilayered analysis with institutional, national, and then international stages—relevant research would be identified at each of these levels and then passed on to the next level for further consideration.

The potential for additional oversight measures for GOF studies of concern to reduce interest in GOF research was raised again. Some participants pointed out that in some cases, such as certain types of research involving human subjects, this was acceptable and appropriate. Gabriel Leung suggested that the longer-term impact would be to discourage scientists from entering into research fields connected to emerging or re-emerging pathogens. This was disputed by others, such as Marc Lipsitch from Harvard University.

Participants also discussed possible reactions by international partners, such as China, if the United States decided to introduce an oversight framework for GOF studies of concern. George Gao felt that China would certainly look closely at such a regime.

5

Summing Up

Harvey Fineberg, chair of the Symposium Planning Committee, explained the plan for the final session. First, he would ask the moderators from the various sessions, all members of the planning committee, to offer their perspectives, summarizing and perhaps adding their own personal comments about key points that were raised in each of the sessions. He then wanted to allow an opportunity for those from the National Institutes of Health (NIH) and the National Science Advisory Board for Biosecurity (NSABB) to raise any issues or topics or questions they would like to address, and to invite further comment from those on the stage and in the audience. Finally, the microphones would be opened for additional comments, suggestions, and ideas that anyone present or listening on the Web would like to include in the record.

Dr. Fineberg invited Charles Haas ("Informing the Policy Framework: The Risk and Benefit Assessment") to give the first summary comments. Dr. Haas began with one editorial comment. Data gaps, particularly on laboratory safety, were thought to limit the ability to do an absolute risk assessment, and there had been a good set of questions from the floor about the need to develop scholarship and support for those studies. His comment was that such data are not totally absent, and it might have been informative to use whatever data were available, even though they were poor, as part of the effort to bound the potential risks that could occur.

Dr. Haas also commented that, if a pure "risk acceptable" rule is to be used as a basis for decision making, it should be recognized that information is lacking on what the level of acceptability should be. Rocco

Casagrande had presented an updated analysis using new data on seasonal versus 1918 influenza, which raised the broader point that risk assessments in general need to be living and to be adaptable to new information as it comes along. Dr. Haas also cited Adam Finkel's statement that leaving uncertainty out is a violation of first principles.

He quoted Dr. Finkel that "Is it safe?" is a vapid question because it is intrinsically without meaning without a reference level. A hierarchy of potential judgment rules exists. Both Tony Cox and Dr. Finkel made that clear, and also that explicit judgments about what rule is to be used need to be made. Kara Morgan called this "deciding how to decide," and she noted that there is rich scholarship from the decision analysis community that needs to be brought to bear. And stakeholder input needs to be included to develop the decision rules.

Dr. Cox had cited the need to avoid the "fallacy of coherence": Just because risk has been accepted in the past does not mean that an informed judgment going forward would make that same numerical risk acceptable. A useful task would be to assess whether or not collection of more information would make a decision better. There is a rich literature on the concept of the value of information in this regard.

Dr. Haas concluded by citing a number of miscellaneous problems that had come up in the discussion. For example, Dr. Casagrande had expressed the concern that bench researchers may not be familiar enough with epidemiological parameters to assess transmissibility. Next, risk–benefit analysis could be used to improve the risk profile of proposed experiments—in other words, envisioning an iterative process of some sort. Dr. Finkel had argued that risk and benefit analyses should be balanced, humble, and explicit about value judgments. And finally, there had been comments from the audience that particularly long-term benefits may be difficult to value and highly uncertain. His editorial comment in response was that, while this may very well be true, it should not mean that one should walk away from the effort to attempt to quantify them using whatever information one had available.

Barry Bloom shared reflections from two sessions, first on behalf of Michelle Mello ("The Policy Landscape: United States") and then from the panel he had moderated ("The Policy Landscape: International Dimensions of Gain-of-Function [GOF] Research"). Dr. Mello's comments included:

- There is no set of policies that targets the specific group of pathogens defined by the NSABB. Instead, the federal policy framework consists of a series of partially overlapping statutes and regulations that are largely tied to specific pathogens and to federal research funding. None of the panelists pointed to major gaps in this framework other than noting that the Department of

Health and Human Services targets a *very* narrow set of experiments and the dual use research of concern (DURC) policy covers only 15 pathogens. However, their comments did reinforce the NSABB's observation that the strength of the policy oversight is stronger for some pathogens than for others.

- Existing law does not really reach research that is not conducted with federal funding (i.e., industry-sponsored research). This raises the question, should it? And if so, through what mechanism?
- The time to regulate is at the time the research is conceived. The point of publication is far too late. Having a strong review process up front avoids a lot of problems down the line—and also establishes that institutions have acted with due care (which may come up in litigation). Funding agencies and institutions can engage principal investigators (PIs) at the point of designing their protocols to think through the risk issues. This is especially useful because many PIs do not understand dual use risk issues.
- Regulators, including both institutions and federal agencies, can benefit from greater use of consultation. Talking with each other and with external experts can boost the quality of review and the dissemination of knowledge and best practices.
- Epistemological question: How do we know if a regulatory approach is working? Beyond the absence of rare, catastrophic events, what should we use as performance measures? The panelists suggested public trust, but in her view, this is both hard to assess and a narrow measure. The NSABB may wish to think (in relation to its Key Finding 2) about what it means to say the policy frameworks are "effective."
- One tension in oversight is between the desire for transparency and the risk that public disclosure of sensitive information will elevate the very dual use risks that oversight is aiming to minimize.
- The criteria that the NSABB set forth for reviewing GOF research are reasonable, but not very specific. They rely on subjective judgments such as "likely" and "highly." Yet there is a tension between pursuing greater specificity in regulations and providing enough flexibility to make case by case judgments. Also, it is not clear how to get more specific about some of these standards.
- How much variation should be tolerated in how institutional review committees evaluate research? On the one hand, one would like to have common standards applied in a reliable fashion. On the other hand, institutions have different capacities, and there might be something one can learn from their individual innovations in practices. The panelists did not see a major prob-

lem with having a "patchwork of institution-dependent rules"; this is something the NSABB may wish to consider.

Dr. Bloom then turned to the comments on the session he had moderated ("The Policy Landscape: International Dimensions of GOF Research"). It was clear from the very beginning of the sessions on the first day that everyone involved in this meeting recognizes that science and the risks and benefits have global implications, and GOF research clearly has raised global concerns. The session included major presentations on the groundbreaking progress made by the European Union (EU), which showed that it was possible to have discussions and bring policies from 28 countries to a common focus, and bring scientific academies in almost all of those countries to a consensus on the scientific policies that would govern this research. The discussions emphasized the need to expand and extend the discussion among countries in Europe. The panelists would be very interested in discussions after the U.S. policies are formulated, and they were eager to find out ways in which discussion and consultation can be expanded to include all countries.

In this context, the session heard a very important discussion of the InterAcademy Partnership, a global network of science and medical academies that now links academies in 128 countries and four regions. That could serve as a useful focus for extending the discussions of GOF research in a coherent way to responsible scientific bodies that already exist and perhaps should be considered in moving forward.

A suggestion that emerged from the session was that the best place to start is probably with discussion within the scientific community rather than going directly to policy makers one at a time, one country at a time, until there is some general understanding and agreement within the scientific community. Then the complexities of those dialogues and discussions could be simplified to a level that could gain understanding and support from political leaders.

The session also heard about the value of not just pontificating but having important partnerships and collaborations that enable transparency, technology transfer, and training to occur. These can also be a way of maintaining standards and identifying low standards that need to be addressed.

He offered several personal reflections about what he had learned during the meeting.

- He had come to the view that process is probably as important as principles. It is not clear, given the technicalities of the science, that the lay public, and even government officials, are going to understand the technicalities. But if the processes at every level are transparent, maybe that is the best way to gain trust within the scientific community and within the public at large. And that means the processes as he was conceiving them, and as the NSABB conceives them, are a set of tiered processes that occurs at multiple levels from the investigator, the Institutional Biosafety Committee (IBC), the institutions, study sections, and all the way up to the higher levels of policy.

- His second reflection on the meeting is that whatever one does, it has to be recognized that science is changing dramatically so that policies cannot be fixed in time to predict what possibilities, opportunities, technologies, and threats will be coming in the future. The policies need to be flexible in some way to accommodate new knowledge and adapt to new opportunities and possibilities and yet have a clear-cut framework that people can work with.
- Finally, he supported Gabriel Leung's comment about why the Biological Weapons Convention (BWC), as far as we know, largely works. Why do the Helsinki principles actually govern how human experimentation is done? He would say it is less legal liability and lawsuits than it is to ask, what are the principal constraints on scientists? He believed those have to do in general with constraints on reputation, credibility, integrity, and respect in the scientific community. Matthew Meselson, for example, when asked how one could possibly encourage more action to enforce the BWC, raised the interesting possibility of making it impossible for scientists who violated international law to travel overseas as another constraint that would be of high value for scientists. So he believed enforcement at a moral level is highly possible.

Baruch Fischhoff offered his comments on the session devoted to "Informing Policy Design: Insights from the Science of Safety and the Science of Public Consultation." He began with some nomenclature, using the term "social science" for those not familiar with that part of the world to include social, behavioral, and decision science. Behavioral science is the study of individuals; it is psychology, microeconomics, neuroscience, and other social sciences. For larger groupings, it is sociology, anthropology, and political science. And decision science is management science, the cost, risk, and benefit analysis of that form of applied mathematics that takes human behavior into consideration. For problems of this complexity and subtlety, he argued that insights from all these fields are needed.

The framing of the human dimensions that he believed came out of the session is that reducing the risks and realizing the benefits of these technologies depends on people at the level of individuals, organizations, and policies. Second, relying on intuition in designing and evaluating the systems that deal with these technologies is natural, but it is unfortunate because those intuitions are often wrong or imprecise. Third, the biological research community faces the challenge of not having what some economists call the absorptive capacity for social science. That is, there is nobody on the inside who can tell when they have a social science problem, define it in terms that would be recognizable to a social scientist, and find somebody who will help them to work on the problem. That is on the demand side. On the supply side, the social science community may lack the incentives for addressing biological science issues because its incentive scheme is to publish on relatively narrow topics. He thought the symposium was fortunate to have speakers in his session who have that bridge which requires them to draw on different social sciences as well as to see the value for the basic science to engage in applied problems.

He then asked what kinds of issues one would find if one brought the social sciences to bear? One is to identify the places in which scientific judgment affects the prediction of outcomes. Many of the statements heard during the symposium had to do with scientists anticipating how transmissible something would be. Given that this a discovery process, there are likely to be surprises. So it is smart to recognize that these are scientific judgments and to elicit them in the best, most accountable way possible. Second, these are ethical judgments and analyses: for example, how you define them, who you share them with, where various publics are engaged in the process. Third is the communication to and from stakeholders so that one can develop the technologies in the ways that are most sensitive to their needs and keep them properly apprised of developments.

A fourth problem, more from the social sciences, is the normalization of pathology and the virtue. One can become accustomed to best practices that are terrible by any absolute standard. But as Ruthanne Huising's talk and Dr. Bloom's comments illustrate, there is also the possibility of the normalization of virtue. There are things that one just does not do, and this is part of the kind of bottom-up process of acculturation and socialization that Dr. Huising discussed.

Fifth, there can be a mismatch between the technology and the regulatory mechanisms in terms of not just government regulation but also the societal controls that one has over technologies. One can have regulatory control mechanisms that do not have the requisite variety for technology that is moving very quickly when institutions were developed for a different environment. Another problem that one runs into is the neglect of opportunity costs. A good deal is known about the technologies in which one has invested and much less about the ones in which one has not invested.

Dr. Fischhoff concluded, in the spirit of Dr. Bloom's two personal comments, with two recommendations.

- Given the difficulty of bridging the basic and social sciences, there would be value in creating centers that would serve as a kind of clearinghouse for helping interested biologists to find social scientists who could help them work with their problems and social scientists to find the people with whom they are willing to work. They could help make the case to department heads that this is a worthy pursuit to spend as much time as all three of the speakers have had working with clients to apply the social science that is available and to create the needed evidence for what some people call adaptive management.
- The second is to develop shadow alternative evaluation processes. That is, if current mechanisms are not up to it, alternative mechanisms are needed. Monica Schoch-Spana's talk illustrated the potential to bound the set of deliberative mechanisms whereby this might work. But one will not really know how they would work until people with the different kinds of expertise and cultural experiences come together and explore them. And one might hope that if there were some worked examples—maybe like some of the conventions that people have talked about—they would eventually become the normal thing that people do. It is very hard to get people to repeal regulations that promise safety, but sometimes they just atrophy. And maybe they will go away if we have something better.

Philip Dormitzer offered his reflections on the ideas raised by what Dr. Fineberg called the "interested parties" in his session ("Best Practices to Inform National Policy Design and Implementation: Perspectives of Key Stakeholders in the Biomedical and Public Health Communities"). He began with Michael Callahan, who pointed out that the European Union and the United States are not the future epicenter—and may not even be the present epicenter—of GOF research. And similarly, government funding may not necessarily be the dominant mode of funding for this research. It is necessary to expand the thinking about how one might influence these processes. Another very interesting point was that some of the case studies he offered where mechanisms of control of infectious agents of concern were lost not due to any malicious intent but due to the necessities facing people operating under difficult circumstances. There are circumstances where consultative mechanisms might help, where forms of assistance might help, and also where incentives need to be created to encourage people to limit risks when there is no capacity to regulate their behavior.

Robert Fisher had discussed the inherent conflict between the need for evidence-based decision making at the regulatory level, which is necessarily time consuming and expensive, and the frequent need to act quickly, particularly in these emerging or outbreak situations. This conflict has to be reconciled, and the considerations around policy for GOF studies of concern play into that. And this also raised the earlier point that estimation of risk can really only be judged in a context of expected benefit. Without benefit, why would one take any risk? These things play into the sorts of mechanisms that one might pursue to try to control the risks of GOF studies of concern.

Dr. Dormitzer commended Jonathan Moreno for trying to identify where there are areas of consensus regarding policy for GOF research. He did not know if everyone agreed on those areas of consensus, but he thought they were close enough to be worth mentioning. There is consensus that there are times when it is necessary to move quickly, but also that some regulation is needed. There is consensus that biocontainment is imperfect, that risk mitigation heavily involves human factors, especially as the mechanical and environmental factors get under better control. He thought that there was consensus it would be desirable to have alternatives to risky experiments, and that gain of function experiments are not fully predictable, but the capacity is probably improving.

Dr. Moreno also had a very interesting proposal for what he called R-BATs, or Risk–Benefit Assessment Teams. The idea is that there would be real-time, ongoing, interactive evaluation of experiments of concern or experiments that may not yet be of concern but could venture into that area so that there was not simply a checkpoint—for example, at the time of funding and another at the time of publication—but an ongoing process of interaction. Dr. Dormitzer thought that might not take care of the whole issue, but it could make a very solid contribution.

Finally, Ethan Settembre had discussed some of the lessons of the first H1N1 pandemic in 2009 and then the H7N9 outbreak response in 2013, making the point that GOF research is an inherent part of the routine business of vaccine production. Unintended consequences of GOF policy choices therefore needed to be considered.

Dr. Dormitzer noted that today sequence analysis is a part of risk analysis and vaccine virus selection, but it is secondary at this point to phenotypic, clinical, and epidemiologic characterizations. He thought, however, that will start to shift over time. It is certainly never going to be the case that a sequence analysis can replace current approaches, but the volume of relevant sequence data is likely to increase dramatically. It is now possible to sequence flu strains directly from harvested secretions; there is no need to grow the virus. The ability to do that sequencing is becoming increasingly widespread, and it is quite conceivable that these will be done in some sort of handheld devices in the coming decade.

Dr. Dormitzer closed with some personal observations. One was an increasing need to consider integration of the multiple biosafety and biosecurity regimens. The other was a concern about unintended consequences: for example, from the "blowback" onto vaccine production from the controversies over GOF studies of concern—or GOF research more generally—in academia.

Ronald Atlas began the discussion of the session he had moderated ("International Governance: Opportunities for Harmonizing GOF Research Policy and Practice") by remarking that he had learned that the international dimensions of the debate about GOF research, risks, and benefits cannot be ignored. A number of possible ways of approaching that on an international scale had been suggested. One was to go to a non-regulatory framework to take ethics or other sorts of systems that have gained traction and are accepted across the biomedical field, build on those, and essentially build a culture of responsibility within the community that would assure the public that everyone was taking the appropriate mitigation steps. Another was to simply accept that nations that were carrying out GOF research would develop their own sets of regulatory frameworks. Another was to allow the efforts that are ongoing in areas like the United States and the European Union to begin to cross-fertilize each other and to bring together groups that would then allow for voluntary harmonization without going to an international organization like the World Health Organization (WHO). And finally, the higher level is to go to a United Nations agency such as WHO and attempt the perhaps impossible task of coming up with a global regulatory scheme.

Dr. Atlas thought that another important point from the session came from Keiji Fukuda: the need to find a compelling and readily understood reason to come together at the international level to take action. What would that reason be for GOF research? Dr. Atlas suggested that it could be "preventing a global pandemic." That could mean that the research is absolutely necessary because it will provide the vaccines, the surveillance, or whatever to prevent the pandemic. Or to take the opposite side, the research itself is a risk because something could get out and cause a pandemic. That is the dilemma underlying the entire debate over GOF research, and he was still not sure there would ever be an answer that was satisfactory to everyone.

Dr. Fineberg then asked if any NSABB members had comments or questions. Joseph Kanabrocki from the University of Chicago and co-chair of the NSABB WG began with some observations. He was heartened that the comments and discussion suggested that the NSABB had not made any major missteps. He was also pleased that there was movement away from a list-based system to a phenotypic system that the NSABB has been recommending for a number of years. That had not been explicitly stated but he thought it was implicit in the discussions.

Dr. Kanabrocki said that, speaking personally, he had heard a number of things on which the NSABB WG had not yet deliberated that he would like to see added to the NSABB report. These included incident reporting mechanisms that could address the lack of data highlighted by the risk and benefit assessment as well as the need for harmonization, both on the national level and the international level. He thought it should be something the final report called for more explicitly, and addressed some of the ideas about how that could be accomplished. He also hoped that the NSABB would recommend a code of conduct for scientists engaged in this type of research.

Dr. Kanabrocki then returned to the three phenotypes recommended in the draft report as the criteria for identifying GOF studies of concern. The original version of the NSABB WG's Draft Working Paper included resistance to countermeasures as an example. He stressed that it was intended only as an example, but, unfortunately, people seemed to have seized on it as the one aspect of the third criterion. So he wanted to remind everyone that for him—and he thought most of the NSABB as well—the third phenotype is what makes this an issue of pandemic potential. He thought the first and second traits go to the animal pathogen interface, and the third trait is where one addresses human public health, the societal aspects of pandemic. He thought that the third trait remains critical, though it might be possible to revise the language in a way that is more palatable.

Susan Wolf, an NSABB member from the University of Minnesota, raised the issue of oversight design and said she wanted to try out two ideas, one at the institutional and one at the federal level. This is crystalized by the flow chart introduced at the symposium (see Figure 2-2). The NSABB has developed the chart to communicate visually the oversight process it is planning. At the institutional level, who decides that an experiment is a potential GOF study of concern? At the moment, the NSABB is envisioning the initial determination would be made by the PI and the local oversight authorities, presumably the IBC. Her concern was how to avoid recapitulating the history of Institutional Review Boards (IRBs), which she characterized as being very slow to design, much less to put in place the sort of "learning" oversight system where there is a systematic effort to gather experience and share lessons learned and also to identify unjustified variations in how the rules are applied. There is a substantial amount of research on this problem and she hoped it would be applied to ensure that the GOF system would be state of the art.

Her other concern was at the federal level and what would happen if a GOF study of concern is identified at the local level. Who would review it and apply the several principles the NSABB was proposing? Could one answer be a new Federal Advisory Committee Act (FACA) committee charged with this task?

Marie-Louise Hammarskjöld, an NSABB member from the University of Virginia, asked about the issue of how to capture research done without federal funding, citing increased interest from industry in university research. She thought that, given that the concern was potential pandemic risk, the board might not be doing its job if it did not deal with that part of the research enterprise.

Jim LeDuc, an NSABB member and Director of the Galveston National Laboratory at the University of Texas Medical Branch, was particularly interested in risk mitigation. His question to the panel was how to create a foundation upon which a policy can be built that clearly articulates the requirements for biosafety and biosecurity, and importantly, a culture of responsibility that spans the scope from the individual scientist all the way through to the institutional leadership.

Dr. Atlas reacted to the question of the IBC versus the national level and suggested that a great deal was learned during the early days of the Recombinant DNA Advisory Committee (RAC). He commented that the IBCs sent cases to the full national board until the RAC was able to demonstrate to the local IBCs what was and was not of greater concern. The RAC refined the principles, and he thought the same approach should be taken for GOF studies of concern. What is needed is to create a learning process, an iterative process, where there is appropriate consultation from the national back to the local and eventually the local learns how to handle the cases and the burden on the national board diminishes.

The RAC had also dealt with the question of federal funding. It turned out that the first cases that came to the RAC were from industry, which wanted the national approval. Industry did not want to go around the system; it wanted to become part of the system even though it was not mandated to do so. He had no reason to think the same thing would not happen here.

Dr. Dormitzer said that he could certainly speak for having been in companies when there are national and accepted standards. Even when not required to follow them, companies in general want to do so. In fact, the most distressing situations are those where there is a lack of clarity over what the expectations are. And that is why the ideas about advisory boards and groups to which companies can turn to ascertain what those standards are, even if compliance is voluntary, are useful. He thought there would be a widespread desire to meet the standards.

Dr. Fineberg added a comment about the discussion of the importance of the scientific community building and reinforcing a culture of safety as well as a discussion about the importance and practicality of public engagement and about the various types of publics. It seemed to him that in the thinking of the NSABB, going forward it would be useful to consider a model that incorporates, at an appropriate level, a FACA-like entity and relevant public participation as a way of building the kind of larger trust, and, frankly, reinforcing the community of safety, both within and around the scientific community, on which success ultimately will depend.

Dr. Fischhoff commented that he was involved with the Food and Drug Administration (FDA) over the past few years as the Center for Drug Evaluation and Research developed a benefit–risk framework (FDA, 2013). The framework was developed jointly with its staff and resembles Kara Morgan's model of deliberative criteria-based frameworks (see Box 2-3). It was designed to help people tell their story in a way that one could see what the logic was; one could compare across decisions; and one could find the decisions that were—as someone has mentioned—anomalous and that gave industry a clearer sense of the kind of things that the FDA was approving.

Dr. Fineberg made another observation on the first and fundamental question of the phenotypic inclusiveness or exclusiveness. One of the things he heard repeatedly in the course of the discussion was the importance of circumscribing the domain of concern so that neither the scientific community nor the regulatory authority, nor, frankly, the interested publics were needlessly burdened with a wide variety of questions that truly do not raise and rise to a level of concern. At the same time, there was a lot of discussion as to whether the current formulation—where the requirement is that a given experiment affects all of the elements—is a sufficient degree of circumscription. He thought that the real challenge for the NSABB was to reflect its actual intent in its description and to do so in a way that is clear and understandable over time. So, for example, he thought that one could be overly fixed on the models that depend on familiarity with influenza as the case. He thought the policy that will be promulgated ultimately needs to be capable of dealing with GOF research, and increasingly, experiments that intend to develop entirely novel organisms with capacities and capabilities that are not currently even expressed in existing microorganisms. And if one thinks that broadly, defining a phenotypic space that involves virulence, and involves transmissibility, and involves resistance to treatment, if that is how one wishes to characterize it, one could imagine placing imaginably any organism at a point in space that has those three attributes defined. Thought of that way, there is an aspect of this space where one would not want research to go at all. There is an aspect of that space where one would not want to require further review. And then there is an aspect of that space, depending on the starting point and the direction of the experiment to make it worse or to make it better—and this is where vaccine development comes in so importantly—would dictate that it may, then, be a topic that requires consideration as a GOF study of concern. He said he hoped that it would be possible for the NSABB to mull over this question and to think about ways to characterize and describe exactly what it believes should determine a consideration for GOF studies of concern. And, perhaps, to be explicit about excluding vaccine development research, which is so fundamental to protection and actually contrary to the concerns. And to be able to apply the principles more generally as new ideas with different organisms will naturally arise in the creative minds of scientists.

Dr. Kanabrocki agreed and said that he wanted to clarify again that, as his NSABB WG co-chair Ken Berns had said on the first day, the NSABB was not really worried about what goes in, but what comes out. The NSABB WG was not saying that the experiments of concern are only those that would result in the three phenotypes. What they were saying is the experiments of concern are those that result in an organism that displays those three phenotypes, and there is a difference. Because one could begin with two of the three and contribute the third and that would be an experiment of concern.

Dr. Fineberg then opened the floor to questions and comments from the participants. Wendy Hall from the Department of Homeland Security asked a question in terms of precedent. First, how important is it that one has full awareness of the GOF experiments being proposed throughout a variety of different labs in the United States? She was not sure there is clarity across the academic community at any one point in time about who is planning and doing what. Her second question related to the experience with the Select Agent rules, which were implemented in 300 various labs with a substantial range in the quality of performance. In GOF research, is there any precedent—if the academic community had full visibility, peer to peer, institution to institution—that there could be corrective elements from the institutional bodies with each other to redirect or help labs not performing as well? Her hope was to avoid the need for the government to have to come down with tough, restrictive language across the board that affects everyone because of a case where one or two labs make an error that makes the mainstream press.

Dr. Fineberg responded that her question reinforced the importance of the scientific community itself coming together in a coherent way on this and related issues of safety and security. From a personal point of view, he did not think the government alone could accomplish this, nor could the community, acting without the guidance of shared standards. So he thought the efforts would be mutually reinforcing.

Dr. Schoch-Spana from the Center for Health Security of the University of Pittsburgh Medical Center picked up a point that Marc Lipsitch from

Harvard University had made about the capacity for innovation, not just prevention. Are there things, such as special research funds, that could incentivize scientists to try alternative approaches to GOF studies of concern? If systems are put in place and data are gathered about the kinds of experiments that are not funded, those data could be synthesized to identify lines of work that need to be replaced with safer alternatives, and research to develop those alternatives could be eligible for special funding.

Nicolas Evans from the University of Pennsylvania offered two comments. The first concerned the Declaration of Helsinki, which was a great initial work in establishing norms in human subjects research and biomedical ethics. But he thought that the FDA's removal of the Declaration of Helsinki from its regulations was an indicator that, as a model for governing the life sciences, one should be especially careful about the way one seeks international collaboration. If the United States sets up or attempts to initiate other arrangements for governing GOF research, only to pull out of them because it does not want them referenced in its own legislation, that would pose a major problem. He also built on Dr. Lipsitch's and Susan Wolf's comments about the critique that IRBs and biomedical ethics chill biomedical research, commenting that it had been made many times and citing two recent works (Klitzman, 2015; Schneider, 2015).

Dr. Evans also offered three other comments.

- He thought it was very important conceptually to make a clear distinction between general GOF research, which is accepted as a valuable and commonly used technique, and specific GOF experiments resulting in the creation of novel pandemic pathogens that is beneficial. For example, the Gryphon Scientific benefits assessment had concluded that a portion of the studies it assessed provided unique benefits.
- Dr. Evans noted that health care workers, the people who bear the disproportionate burden of risk in the event of an infectious disease outbreak, had been entirely absent from the discussions.
- Regarding innovation, he commented that because $820 million had been provided to synthetic biology research over the past half decade, it seemed prudent to also spend a small amount of money on innovation in applied biosafety, such as on material science to improve personal protective equipment.

Jenna Ogilvie from the National Academies of Sciences, Engineering, and Medicine's staff brought two questions from the Web. The first was from Grigory Khimulya from Harvard College. Do current oversight frameworks provide adequate treatment of novel pathogens that were never seen before and are not on the pathogen lists mentioned in the NSABB's draft recommendations? For example, if a new potentially pandemic pathogen like Middle East respiratory syndrome (MERS) is identified, would GOF studies of concern with this pathogen fall under proposed regulation? The second question, to Dr. Casagrande from Gryphon Scientific, came from John Kadvany from Policy & Decision Science in Menlo Park, California, prompted by publications suggesting that GOF research has characteristics of so-called potential "normal accidents," in which a technology combines highly negative outcomes (e.g., a nuclear plant meltdown) with unquantified and perhaps unquantifiable scenarios falling outside even the most complete probabilistic risk analysis. Gryphon Scientific's work suggests that such scenarios may be relevant with the extreme negative outcome being pandemic risk. Did Dr. Casagrande have an opinion on this characterization of GOF studies of concern? Is it correct in some respects as it may be for some contemporary technologies? Or is there a characterization fueling clashing GOF risk perceptions?

Dr. Casagrande commented from outside his role as PI of the risk and benefit assessment to push back a little bit on several comments he had heard about what could be learned from the successes of the BWC. He thought that the protocol was a better exemplar because it banned first use of bacteriological warfare. In contrast, several members of the BWC have violated its provisions, leading him to conclude that one ought to learn from its failures, such as the lack of a verification and inspection regime and the lack of an enforcement capability that is relevant internationally.

Dr. Lipsitch from Harvard University commented that there had been considerable discussion about whether there is consensus that there are any experiments that everyone would agree would never be acceptable and any experiments everyone would agree should never be impeded. He said he could certainly think of experiments and developments one would never want to impede and suggested that there should be a green line as well as a red line. He thought that whatever regulatory framework or oversight framework is developed, it would be incredibly helpful to have at least those two kinds of cases spelled out by some examples in order to build our intuition for the next time something comes up that is not envisioned yet. He also thought some more contestable case studies, where there would not be an easy consensus, would be useful.

Dr. Kanabrocki responded to Dr. Lipsitch. The NSABB WG had tried on a number of occasions to think of experiments that absolutely should not be done. And every single example that came up was of an experiment that lacked scientific merit. So he suggested that, in his personal view, it would be a struggle to think of experiments that have scientific merit that should not be done.

Gerald Epstein from the Department of Homeland Security suggested that it would be useful to go back to the Department of Health and Human Services (HHS) framework that Larry Kerr had described on the first day, and the test that a proposed project would have to satisfy before it was deemed acceptable for funding. One was that the pathogen to be constructed was one that might occur by a natural process, so that there was a reasonable expectation nature might get there first. If it is not something nature might do on its own, one could not argue the work was to defend against a potential natural development. This might be an example of something on the other side of the line, at least from the precedent of the existing HHS framework.

Dr. Fineberg closed the session by expressing the Academies' deep appreciation to everyone who had taken part, in person or via the Web. He commended the work being done in Europe and commented that, in his view, a policy about GOF research that applies only to one country is not a policy that will work for the safety of the world. And that is something of which one needed to be very mindful. He also commented that it was evident from all the discussion that whatever the next iteration of conclusions and recommendations that emerge from the NSABB is, it will really be one step in a process that is likely to continue. It will require continued refinement, the engagement of the scientific community, and finding creative ways for the public that is interested and affected by GOF research to be involved in the process of decision making going forward.

Bibliography

AL-Hmoud, N. 2016. Comments. Presentation at Gain-of-Function—The Second Symposium, National Academies of Sciences, Engineering, and Medicine, Washington, DC, March 11. Available at: https://www.youtube.com/playlist?list=PLuTGMA3A_-175RgOxOwYyy_1A6gBHZTYN.

Bates, S., and J. Holroyd. 2012. Human Factors That Lead to Non-Compliance with Standard Operating Procedures. Health and Safety Executive Research Report RR919. Available at: http://www.hse.gov.uk/research/rrpdf/rr919.pdf.

Callahan, M. 2016. Exportation of GOF Policy to International Stakeholders: Case Studies from Recent Events. Presentation at Gain-of-Function—The Second Symposium, National Academies of Sciences, Engineering, and Medicine, Washington, DC, March 11. Available at: https://www.youtube.com/playlist?list=PLuTGMA3A_-175RgOxOwYyy_1A6gBHZTYN.

Casagrande, R. 2016. Risk and Benefit Analysis (RBA) of Gain-of-Function Research: Gaps and Future Considerations. Presentation at Gain-of-Function—The Second Symposium, National Academies of Sciences, Engineering, and Medicine, Washington, DC, March 10. Available at: https://www.youtube.com/playlist?list=PLuTGMA3A_-175RgOxOwYyy_1A6gBHZTYN.

CDC and NIH (Centers for Disease Control and Prevention and National Institutes of Health). 2007. *Biosafety in Microbiological and Biomedical Laboratories*, 5th ed. (L. Casey Chosewood and Deborah E. Wilson, eds.). Washington, DC: U.S. Government Printing Office.

Chandler, F., I. A. Heard, M. Presley, A. Burg, E. Mideen, and P. Mongon, P. 2010. NASA Human Error Analysis. *National Aeronautics and Space Administration, Final Report*. Washington, DC: NASA.

Coglianese, C. 2015. *Listening, Learning, Leading: A Framework for Regulatory Excellence*. Phiadelphia, PA: Penn Program on Regulation. Available at: https://www.law.upenn.edu/live/files/4946-pprfinalconvenersreport.pdf.

Commission on a Global Health Risk Framework for the Future. 2015. *The Neglected Dimension of Global Security: A Framework to Counter Infectious Disease Crises.* doi: 10.17226/21891.

Cotton-Barratt, O., S. Farquhar, and A. Snyder-Beattie. 2016. Beyond Risk-Benefit Analysis: Pricing Externalities for Gain-of-Function Research of Concern. Policy Working Paper [Revision 0.9]. Future of Humanity Institute, University of Oxford. Available at: http://globalprioritiesproject.org/wp-content/uploads/2016/03/GoFv9-3.pdf.

Cox, T. 2016. Using Risk Benefit Analysis to Improve GOF Research Decisions. Presentation at Gain-of-Function—The Second Symposium, National Academies of Sciences, Engineering, and Medicine, Washington, DC, March 10. Available at: https://www.youtube.com/playlist?list=PLuTGMA3A_-175RgOxOwYyy_1A6gBHZTYN.

Davidson, E. M., R. Frothingham, and R. Cook-Deegan. 2007. Practical Experiences in Dual-Use Review. *Science* 316:1432-1433.

Draghia-Akli, R. 2016. Gain of Function/Dual Use: EU Perspective. Presentation at Gain-of-Function—The Second Symposium, National Academies of Sciences, Engineering, and Medicine, Washington, DC, March 10. Available at: https://www.youtube.com/playlist?list=PLuTGMA3A_-175RgOxOwYyy_1A6gBHZTYN.

EASAC (European Academies Scientific Advisory Council). 2015a. *Gain-of-Function: Experimental Applications Relating to Potentially Pandemic Pathogens*. EASAC Policy Report 27. Halle (Saale), Germany: EASAC. Available at: http://www.easac.eu/fileadmin/PDF_s/reports_statements/Gain_of_Function/EASAC_GOF_Web_complete_centred.pdf.

EASAC. 2015b. Summary of EASAC Launch Event on Gain of Function. Available at: http://www.easac.eu/fileadmin/PDF_s/reports_statements/Gain_of_Function/Summary_of_EASAC_launch_event_on_Gain_of_Function_FINAL.pdf.

Epstein, G. 2016. NSABB Working Group Proposal. Presentation at Gain-of-Function—The Second Symposium, National Academies of Sciences, Engineering, and Medicine, Washington, DC, March 10. Available at: https://www.youtube.com/playlist?list=PLuTGMA3A_-175RgOxOwYyy_1A6gBHZTYN.

European Society for Virology. 2013. Letter to the President of the European Commission. October 16. Available at: http://wp.eusv.eu/wp-content/uploads/2015/10/ESV-letter-on-Gain-of-function_GOF_research-in-Virology.pdf.

FDA (Food and Drug Administration). 2013. *Structured Approach to Benefit-Risk Assessment in Drug Regulatory Decision-Making*. PDUFA V Plan (FY 2013-2017). Available at: http://www.fda.gov/downloads/ForIndustry/UserFees/PrescriptionDrugUserFee/UCM329758.pdf.

Fineberg, H. 2016. Presentation at Gain-of-Function—The Second Symposium, National Academies of Sciences, Engineering, and Medicine, Washington, DC, March 10. Available at: https://www.youtube.com/playlist?list=PLuTGMA3A_-175RgOxOwYyy_1A6gBHZTYN.

Finkel, A. M. 2016. Solution-Focused Risk/Benefit Assessment (RBA) for Gain-of-Function Research. Presentation at Gain-of-Function—The Second Symposium, National Academies of Sciences, Engineering, and Medicine, Washington, DC, March 10. Available at: https://www.youtube.com/playlist?list=PLuTGMA3A_-175RgOxOwYyy_1A6gBHZTYN.

Fisher, R. 2016. A Regulator's Perspective. Presentation at Gain-of-Function—The Second Symposium, National Academies of Sciences, Engineering, and Medicine, Washington, DC, March 11. Available at: https://www.youtube.com/playlist?list=PLuTGMA3A_-175RgOxOwYyy_1A6gBHZTYN.

Foundation for Vaccine Research. 2013. Letter to the President of the European Commission. December 18. Available at: http://www.nature.com/polopoly_fs/7.14586!/file/vaccine%20foundation%20letter.pdf.

Frothingham, R. 2016. Lesson Learned from Thirteen Years of Gain-of-Function Research Reviews. Presentation at Gain-of-Function—The Second Symposium, National Academies of Sciences, Engineering, and Medicine, Washington, DC, March 10. Available at: https://www.youtube.com/playlist?list=PLuTGMA3A_-175RgOxOwYyy_1A6gBHZTYN.

Fukuda, K. 2016. International Policy Landscape. Presentation at Gain-of-Function—The Second Symposium, National Academies of Sciences, Engineering, and Medicine, Washington, DC, March 10. Available at: https://www.youtube.com/playlist?list=PLuTGMA3A_-175RgOxOwYyy_1A6gBHZTYN.

GAO (Government Accountability Office). 2009. *High-Containment Laboratories: National Strategy for Oversight Is Needed*. GAO-09-574. Washington, DC: U.S. Government Printing Office.

Gao, G. F. 2016. Comments. Presentation at Gain-of-Function—The Second Symposium, National Academies of Sciences, Engineering, and Medicine, Washington, DC, March 11. Available at: https://www.youtube.com/playlist?list=PLuTGMA3A_-175RgOxOwYyy_1A6gBHZTYN.

GEC (German Ethics Council). 2014. *Biosecurity: Freedom and Responsibility of Research*. Opinion. Berlin: Deutscher Ethikrat. Available at: http://www.ethikrat.org/files/opinion-biosecurity.pdf.

Gryphon Scientific. 2015. *Risk and Benefit Analysis of Gain of Function Research*. Draft Final Report. Takoma Park, MD: Gryphon Scientific. Available at: http://osp.od.nih.gov/sites/default/files/Risk%20and%20Benefit%20Analysis%20of%20Gain%20of%20Function%20Research%20-%20Draft%20Final%20Report.pdf.

Gryphon Scientific. 2016. *Risk and Benefit Analysis of Gain of Function Research: Final Report—April 2016*. Takoma Park, MD: Gryphon Scientific. Available at: http://www.gryphonscientific.com/wp-content/uploads/2016/04/Risk-and-Benefit-Analysis-of-Gain-of-Function-Research-Final-Report.pdf.

Handelsman, J. 2016. Opening Remarks. Presentation at Gain-of-Function—The Second Symposium, National Academies of Sciences, Engineering, and Medicine, Washington, DC, March 10. Available at: https://www.youtube.com/playlist?list=PLuTGMA3A_-175RgOxOwYyy_1A6gBHZTYN.

Haynes, A. B., T. G. Weiser, W. R. Berry, S. R. Lipsitz, A-H. S. Breizat, E. P. Dellinger, T. Herbosa, S. Joseph, P. L. Kibatala, M. C. M. Lapitan, A. F. Merry, K. Moorthy, R. K. Reznick, B. Taylor, and A. A. Gawande. 2009. A surgical safety checklist to reduce morbidity and mortality in a global population. *New England Journal of Medicine* 360(5):491-499.

HHS (Department of Health and Human Services). 2012. *A Framework for Guiding U.S. Department of Health and Human Services Funding Decisions About Research Proposals with the Potential for Generating Highly Pathogenic Avian Influenza H5N1 Viruses That Are Transmissible Among Mammals by Respiratory Droplets*. Available at: http://www.phe.gov/s3/dualuse/Documents/funding-hpai-h5n1.pdf.

Holdren, J. P., and L. Monaco. 2014. Enhancing Biosafety and Biosecurity in the United States. August 18. Available at: http://www.whitehouse.gov/sites/default/files/microsites/ostp/enhancing_biosafety_and_biosecurity_19aug2014_final.pdf.

Huising, R. 2016. Laboratory Safety and Security: Organizational and Cultural Factors. Presentation at Gain-of-Function—The Second Symposium, National Academies of Sciences, Engineering, and Medicine, Washington, DC, March 11. Available at: https://www.youtube.com/playlist?list=PLuTGMA3A_-175RgOxOwYyy_1A6gBHZTYN.

Huising, R., and S. S. Silbey. 2011. Governing the Gap: Forging Safe Science Through Relational Regulation. *Regulation & Governance* 5(1):14-42.

Huntley-Fenner, G. 2016. The Lack of Human Reliability Data Is a Barrier to the Attainment of the NSABB's Risk Reduction Objectives. Presentation at Gain-of-Function—The Second Symposium, National Academies of Sciences, Engineering, and Medicine, Washington, DC, March 11. Available at: https://www.youtube.com/playlist?list=PLuTGMA3A_-175RgOxOwYyy_1A6gBHZTYN.

IDSA (Infectious Disease Society of America). 2016. Comments to the NSABB Working Paper on Evaluating the Risks and Benefits of Gain-of-Function Studies to Formulate Policy Recommendations. Submitted to NSABB February 23. Available at: http://www.idsociety.org/uploadedFiles/IDSA/Policy_and_Advocacy/Current_Topics_and_Issues/Support_for_Medical_Education_and_Research/Letters/IDSA%20Comments%20to%20the%20NSABB%2002222016.pdf.

Kerr, L. 2016. The HHS Framework for Review of Certain Avian Influenza Experiments. Presentation at Gain-of-Function—The Second Symposium, National Academies of Sciences, Engineering, and Medicine, Washington, DC, March 10. Available at: https://www.youtube.com/playlist?list=PLuTGMA3A_-175RgOxOwYyy_1A6gBHZTYN.

Klitzman, R. 2015. *The Ethics Police?: The Struggle to Make Human Research Safe*. New York: Oxford University Press.

Leung. G. 2016. Comments. Presentation at Gain-of-Function—The Second Symposium, National Academies of Sciences, Engineering, and Medicine, Washington, DC, March 11. Available at: https://www.youtube.com/playlist?list=PLuTGMA3A_-175RgOxOwYyy_1A6gBHZTYN.

Lipsitch, M., D. A. Relman, and T. V. Inglesby. 2016. COMMENTARY: Six policy options for conducting gain-of-function research. *CIDRAP Newsletter*. March 8. Available at: http://www.cidrap.umn.edu/news-perspective/2016/03/commentary-six-policy-options-conducting gain-function-research.

Moreno, J. 2016. Presentation. Presentation at Gain-of-Function—The Second Symposium, National Academies of Sciences, Engineering, and Medicine, Washington, DC, March 11. Available at: https://www.youtube.com/playlist?list=PLuTGMA3A_-175RgOxOwYyy_1A6gBHZTYN.

Morgan, K. 2016. Informing the Policy Framework. Presentation at Gain-of-Function—The Second Symposium, National Academies of Sciences, Engineering, and Medicine, Washington, DC, March 10. Available at: https://www.youtube.com/playlist?list=PLuTGMA3A_-175RgOxOwYyy_1A6gBHZTYN.

NIH (National Institutes of Health). 2013. *NIH Guidelines for Research Involving Recombinant or Synthetic Nucleic Acid Molecules*. Available at: http://osp.od.nih.gov/office-biotechnology-activities/biosafety/nih-guidelines.

NRC (National Research Council). 2004. *Biotechnology Research in an Age of Terrorism*. Washington, DC: The National Academies Press.

NRC. 2015. *Potential Risks and Benefits of Gain-of-Function Research: Summary of a Workshop*. Washington, DC: The National Academies Press.

NSABB (National Science Advisory Board for Biosecurity). 2015a. *Working Paper Prepared by the NSABB Working Group on Evaluating the Risks and Benefits of Gain-of-Function Studies to Formulate Policy Recommendations*. December 23. Available at: http://osp.od.nih.gov/sites/default/files/NSABB%20WG%20Working%20Paper%20on%20Gain-of-Function%20Studies%2012-23-2015_0.pdf.

NSABB. 2015b. *Framework for Conducting Risk and Benefit Assessments of Gain-of-Function Research: Recommendations of the National Science Advisory Board for Biosecurity*. Available at: http://osp.od.nih.gov/sites/default/files/resources/NSABB_Framework_for_Risk_and_Benefit_Assessments_of_GOF_Research-APPROVED.pdf.

Potter, P. 2016. Presentation at Gain-of-Function—The Second Symposium, National Academies of Sciences, Engineering, and Medicine, Washington, DC, March 10. Available at: https://www.youtube.com/playlist?list=PLuTGMA3A_-175RgOxOwYyy_1A6gBHZTYN.

Schneider, C. E. 2015. *The Censor's Hand: The Misregulation of Human-Subject Research*. Cambridge, MA: MIT Press.

Schoch-Spana, M. 2016. Public Deliberation and Gain-of-Function Research Policy: Putting It into Practice. Presentation at Gain-of-Function—The Second Symposium, National Academies of Sciences, Engineering, and Medicine, Washington, DC, March 11. Available at: https://www.youtube.com/playlist?list=PLuTGMA3A_-175RgOxOwYyy_1A6gBHZTYN.

Selgelid, M. J. 2015. *White Paper: Gain-of-Function Research: Ethical Analysis*. Available at: http://osp.od.nih.gov/sites/default/files/Gain-of-Function%20Research%20Ethical%20Analysis%20White%20Paper%20by%20Michael%20Selgelid_0.pdf.

Selgelid, M. J. 2016. Comments. Presentation at Gain-of-Function—The Second Symposium, National Academies of Sciences, Engineering, and Medicine, Washington, DC, March 11. Available at: https://www.youtube.com/playlist?list=PLuTGMA3A_-175RgOxOwYyy_1A6gBHZTYN.

Settembre, E. 2016. Influenza Vaccine Production Considerations. Presentation at Gain-of-Function—The Second Symposium, National Academies of Sciences, Engineering, and Medicine, Washington, DC, March 11. Available at: https://www.youtube.com/playlist?list=PLuTGMA3A_-175RgOxOwYyy_1A6gBHZTYN.

Stanley, S. L., Jr. 2016. NSABB Report: Preliminary Findings and Draft Recommendations About Gain-of-Function Research. Presentation and Discussion at Gain-of-Function—The Second Symposium, National Academies of Sciences, Engineering, and Medicine, Washington, DC, March 10. Available at: https://www.youtube.com/playlist?list=PLuTGMA3A_-175RgOxOwYyy_1A6gBHZTYN.

ter Meulen, V. 2016. Gain of Function Research: Advice from EASAC. Presentation at Gain-of-Function—The Second Symposium, National Academies of Sciences, Engineering, and Medicine, Washington, DC, March 10. Available at: https://www.youtube.com/playlist?list=PLuTGMA3A_-175RgOxOwYyy_1A6gBHZTYN.

U.S. Government. 2015. *Report of the Federal Experts Security Advisory Panel*. Available at: http://www.phe.gov/s3/Documents/fesap.pdf.

Vöneky, S. 2016. The German Ethics Council Report. Presentation at Gain-of-Function—The Second Symposium, National Academies of Sciences, Engineering, and Medicine, Washington, DC, March 10. Available at: https://www.youtube.com/playlist?list=PLuTGMA3A_-175RgOxOwYyy_1A6gBHZTYN.

White House. 2012. *USG Policy for Oversight of Life Sciences Dual Use Research of Concern*. March 29. Available at: http://www.phe.gov/s3/dualuse/Documents/us-policy-durc-032812.pdf.

White House. 2014a. *U.S. Government Gain-of-Function Deliberative Process and Research Funding Pause on Selected Gain-of-Function Research Involving Influenza, MERS, and SARS Viruses*. October 17. Available at: http://www.phe.gov/s3/dualuse/Documents/gain-of-function.pdf.

White House. 2014b. *USG Policy for Institutional Oversight of Life Sciences Dual Use Research of Concern*. September 24. Available at: http://www.phe.gov/s3/dualuse/Documents/durc-policy.pdf.

WHO (World Health Organization). 2006. *Biorisk Management: Laboratory Biosecurity Guidance*. Geneva: World Health Organization.

WHO. 2010. *Responsible Life Sciences Research for Global Health Security: A Guidance Document*. Geneva: World Health Organization.

Willemarck, N., B. Brosius, B. Van Vaerenbergh, A. Leunda, A. Baldo, and C. D. D. Thi. 2012. *Laboratory-Acquired Infections in Flanders (2007-2012): An Online Survey*. Brussels: Institut Scientifique de Santé Publique.

Wolinetz, C. D. 2016. The Gain-of-Function Deliberative Process. Presentation at Gain-of-Function—The Second Symposium, National Academies of Sciences, Engineering, and Medicine, Washington, DC, March 10. Available at: https://www.youtube.com/playlist?list=PLuTGMA3A_-175RgOxOwYyy_1A6gBHZTYN.

Appendix A

Board and Committee Members of Collaborating Units

BOARD ON LIFE SCIENCES

Chair

JAMES P. COLLINS (*Chair*), Virginia M. Ullman Professor of Natural History and the Environment, School of Life Sciences, College of Liberal Arts and Sciences, Arizona State University

Members

ENRIQUETA C. BOND, President Emeritus, Burroughs Wellcome Fund
ROGER D. CONE, Professor and Chairman of Molecular Physiology and Biophysics, Vanderbilt University Medical Center
NANCY D. CONNELL, Professor of Medicine and Director, Center for the Study of Emerging and Re-emerging Pathogens, New Jersey Medical School
JOSEPH R. ECKER, Professor, Salk International Council Chair in Genetics, Salk Institute for Biological Studies
SARAH C. R. ELGIN, Professor of Biology, Department of Biology, Washington University
LINDA G. GRIFFITH, Professor of Biological Engineering, Massachusetts Institute of Technology
ELIZABETH HEITMAN, Associate Professor, Center for Biomedical Ethics and Society, Vanderbilt University Medical Center
RICHARD A. JOHNSON, CEO, Global Helix LLC

JUDITH KIMBLE, Vilas Professor of Biochemistry and Molecular Biology and Medical Genetics, University of Wisconsin, Madison
MARY E. MAXON, Biosciences Principal Deputy, Lawrence Berkeley National Laboratory
JILL P. MESIROV, Professor, Department of Medicine, San Diego School of Medicine and Moores Cancer Center
KAREN E. NELSON, President, J. Craig Venter Institute
CLAIRE POMEROY, President, Albert and Mary Lasker Foundation
MARY E. POWER, Professor, Department of Integrative Biology, University of California, Berkeley
MARGARET RILEY, President, Massachusetts Academy of Sciences; Professor of Biology, University of Massachusetts
LANA SKIRBOLL, Vice President, Academic and Scientific Affairs, Sanofi
JANIS WEEKS, Professor, Department of Biology, University of Oregon

Staff

FRANCES SHARPLES, Director
LIDA ANESTIDOU, Senior Program Officer
KATIE BOWMAN, Senior Program Officer
JO HUSBANDS, Senior Scholar
JAY LABOV, Senior Scholar
KEEGAN SAWYER, Program Officer
MARILEE SHELTON-DAVENPORT, Senior Program Officer
AUDREY THEVENON, Associate Program Officer
BETHELHEM MEKASHA, Financial Associate
VANESSA LESTER, Research Associate
ANGELA KOLESNIKOVA, Administrative Assistant
JENNA OGILVIE, Senior Program Assistant
AANIKA SENN, Senior Program Assistant

BOARD ON HEALTH SCIENCES POLICY

Chair

JEFFREY KAHN (*Chair*), Levi Professor of Bioethics and Public Policy, Johns Hopkins University

Members

WYLIE BURKE, Professor and Chair, Bioethics and Humanities Department, University of Washington
R. ALTA CHARO, Warren P. Knowles Professor of Law and Bioethics, University of Wisconsin, Madison
LINDA H. CLEVER, Senior Physician, Occupational Health, California Pacific Medical Center; President, RENEW
BARRY S. COLLER, Physician in Chief, Vice President for Medical Affairs, and David Rockefeller Professor, The Rockefeller University
BERNARD A. HARRIS, CEO and Managing Director, Vesalius Ventures
MARTHA N. HILL, Dean Emerita and Professor of Nursing, Medicine and Public Health, Johns Hopkins University School of Nursing
STEVEN E. HYMAN, Director, Stanley Center, The Broad Institute
ALAN M. JETTE, Professor of Health Policy & Management, Boston University School of Public Health
PATRICIA A. KING, Carmack Waterhouse Professor of Law, Medicine, Ethics and Public Policy, Georgetown University Law Center
STORY C. LANDIS, Scientist Emeritus, National Institute of Neurological Disorders and Stroke
HARRY T. ORR, Director, Institute of Human Genetics, University of Minnesota
UMAIR SHAH, Executive Director, Harris County Public Health & Environmental Services
ROBYN STONE, Senior Vice President of Research, LeadingAge
SHARON TERRY, President and CEO, Genetic Alliance

Board Director

ANDREW M. POPE, Senior Board Director

COMMITTEE ON SCIENCE, TECHNOLOGY, AND LAW

Co-Chairs

DAVID BALTIMORE, President Emeritus and Robert Andrews Millikan Professor of Biology, California Institute of Technology

DAVID S. TATEL, Judge, U.S. Court of Appeals for the District of Columbia Circuit

Members

THOMAS D. ALBRIGHT, Conrad T. Prebys Chair in Vision Research, Salk Institute for Biological Studies

ANN ARVIN, Lucile Packard Professor of Pediatrics and Microbiology and Immunology; Vice Provost and Dean of Research, Stanford University

BARBARA E. BIERER, Faculty Co-chair, Multi-Regional Clinical Trials Center, Harvard University (Harvard MRCT), and Professor of Medicine, Harvard Medical School and Brigham and Women's Hospital

CLAUDE CANIZARES, Bruno Rossi Professor of Physics, Massachusetts Institute of Technology

ARTURO CASADEVALL, Professor and Chair, W. Harry Feinstone Department of Molecular Microbiology and Immunology, Johns Hopkins Bloomberg School of Public Health

JOE S. CECIL, Project Director, Program on Scientific and Technical Evidence, Division of Research, Federal Judicial Center

R. ALTA CHARO, Warren P. Knowles Professor of Law and Bioethics, University of Wisconsin, Madison

HARRY T. EDWARDS, Judge, U.S. Court of Appeals for the District of Columbia Circuit

DREW ENDY, Associate Professor, Bioengineering, Stanford University, and President, BioBricks Foundation

MARCUS FELDMAN, Burnet C. and Mildred Wohlford Professor of Biological Sciences, Stanford University

JEREMY FOGEL, Director, Federal Judicial Center

HENRY T. GREELY, Deane F. and Kate Edelman Johnson Professor of Law and Professor, by courtesy, of Genetics, Stanford School of Medicine

MICHAEL GREENBERGER, Law School Professor and Director, Center for Health and Homeland Security, University of Maryland

MICHAEL IMPERIALE, Arthur F. Thurnau Professor of Microbiology and Immunology, University of Michigan

GREG KISER, Chief Technologist, Intellectual Ventures

ROBERT S. LANGER, David H. Koch Institute Professor, Massachusetts Institute of Technology
GOODWIN, LIU, Associate Justice, California Supreme Court
JENNIFER MNOOKIN, Dean, and David G. Price and Dallas P. Price Professor of Law, University of California, Los Angeles, School of Law
R. GREGORY MORGAN, Senior Vice President and Secretary of the Corporation, Massachusetts Institute of Technology
HARRIET RABB, Vice President and General Counsel, The Rockefeller University
DAVID A. RELMAN, Thomas C. and Joan M. Merigan Professor, Departments of Medicine, and of Microbiology and Immunology, Stanford University and Chief, Infectious Disease Section, Veterans Affairs Palo Alto Health Care System
MARTINE A. ROTHBLATT, Chairman and Chief Executive Officer, United Therapeutics
JOSHUA R. SANES, Professor of Molecular and Cellular Biology and Paul J. Finnegan Family Director, Center for Brain Science, Harvard University
DAVID VLADECK, Professor and Co-Director, Institute for Public Representation, Georgetown Law School

Staff

ANNE-MARIE MAZZA, Director
STEVEN KENDALL, Program Officer
KAROLINA KONARZWESKA, Program Coordinator

Appendix B

Committee Biographies

Harvey V. Fineberg (*Committee Chair*) is the president of the Gordon and Betty Moore Foundation and served two consecutive terms as president of the Institute of Medicine (IOM) (2002-2014), now known as the National Academy of Medicine. He served as provost of Harvard University from 1997 to 2001, following 13 years as dean of the Harvard School of Public Health. He has devoted most of his academic career to the fields of health policy and medical decision making. His past research has focused on the process of policy development and implementation, assessment of medical technology, evaluation and use of vaccines, and dissemination of medical innovations. Dr. Fineberg helped found and served as president of the Society for Medical Decision Making and has been a consultant to the World Health Organization. At the IOM, he chaired and served on a number of panels dealing with health policy issues, ranging from AIDS to new medical technology. He also served as a member of the Public Health Council of Massachusetts (1976-1979), as chairman of the Health Care Technology Study Section of the National Center for Health Services Research (1982-1985), and as president of the Association of Schools of Public Health (1995-1996). Dr. Fineberg serves on the board of the Hewlett Foundation and chairs the board of the Carnegie Endowment for International Peace. Dr. Fineberg is co-author of the books *Clinical Decision Analysis*, *Innovators in Physician Education*, and *The Epidemic that Never Was*, an analysis of the controversial federal immunization program against swine flu in 1976. He has co-edited several books on such diverse topics as AIDS prevention, vaccine safety, global health, and understanding risk in society. He has also authored numerous articles published in professional journals. Dr. Fineberg is the recipient of several honorary degrees and the Stephen Smith Medal for Distinguished Contributions in Public Health from the New York Academy of Medicine. He earned his bachelor's and doctoral degrees from Harvard University.

Ronald M. Atlas is professor of biology at the University of Louisville. After receiving his master's and doctoral degrees from Rutgers University, he became a postdoctoral fellow at the Jet Propulsion Laboratory where he worked on Mars life detection. He has served as chair of the National Aeronautics and Space Administration's Planetary Protection Subcommittee, co-chair of the American Society for Microbiology (ASM) Task Force on Biodefense, and a member of the Federal Bureau of Investigation Scientific Working Group on Microbial Genetics and Forensics. He also served as president of ASM and was a member of the National Institutes of Health Recombinant Advisory Committee. He currently chairs the Public and Scientific Affairs Board of the ASM. His research has included development of detection methods for pathogens in the environment. Dr. Atlas is author of nearly 300 manuscripts and 20 books, and regularly advises the U.S. government on policy issues related to the deterrence of bioterrorism.

Ruth L. Berkelman is the Rollins Chair and Director of the Center for Public Health Preparedness and Research at the Rollins School of Public Health at Emory University. She holds appointments in the departments of epidemiology, global health and medicine, and serves as a senior associate faculty member in Emory's Center for Ethics. She previously served as an assistant surgeon general in the U.S. Public Health Service at the Centers for Disease Control and Prevention (CDC). Elected to the Institute of Medicine (now the National Academy of Medicine) in 2004, she has served on various committees, including the Forum on Emerging Infectious Diseases and the Board on Life Sciences. She has been a member of the National Biodefense Science Board and the Board of Trustees at Princeton University. She was previously Chair of the Public and Scientific Affairs Board of the American Society of Microbiology. She currently chairs the Board of Scientific Counselors for infectious diseases at CDC.

Barry R. Bloom is a leading scientist in the areas of infectious diseases, vaccines, and global health and is a former consultant to the White House. Dr. Bloom enjoyed a distinguished career in bench science as the principal investigator of a laboratory researching the immune response to tuberculosis. He has been extensively involved with the World Health Organization (WHO) for more than 40 years. He was Chair of the Technical and Research Advisory Committee to the Global Programme on

Malaria at WHO and a member of the WHO Advisory Committee on Health Research, as well as chairing the WHO Committees on Leprosy Research and Tuberculosis Research, and the Scientific and Technical Advisory Committee of the United Nations Development Programme/World Bank/WHO Special Programme for Research and Training in Tropical Diseases. Dr. Bloom serves on the editorial board of the *Bulletin of the World Health Organization*. Dr. Bloom served on the Ellison Medical Foundation Scientific Advisory Board and the Wellcome Trust Pathogens, Immunology and Population Health Strategy Committee. He was on the Scientific Advisory Board of the Earth Institute at Columbia University and the Advisory Council of the Paul G. Rogers Society for Global Health Research. His past service includes membership on the National Advisory Council of the National Institute for Allergy and Infectious Diseases, the Scientific Advisory Board of the National Center for Infectious Diseases of the Centers for Disease Control and Prevention, and the National Advisory Board of the Fogarty International Center at the National Institutes of Health, as well as the Governing Board of the Institute of Medicine, now known as the National Academy of Medicine. Dr. Bloom was the founding chair of the board of trustees for the International Vaccine Institute in South Korea. He has chaired the Vaccine Advisory Committee of UNAIDS where he played a critical role in the debate surrounding the ethics of AIDS vaccine trials. He was also a member of the U.S. AIDS Research Committee. Dr. Bloom was introduced to the Harvard Chan School as the Dean of Faculty in 1998, and stepped down December 31, 2008. Dr. Bloom now serves as a Harvard University Distinguished Service Professor.

Donald S. Burke is the dean of the Graduate School of Public Health, director of the Center for Vaccine Research, and associate vice chancellor for global health at the University of Pittsburgh. He is also the first occupant of the University of Pittsburgh Medical Center-Jonas Salk Chair in Global Health and a distinguished university professor of health science and policy. He was an intern and resident in medicine at Boston City and Massachusetts General Hospitals and trained as a research fellow in infectious diseases at the Walter Reed Army Medical Center. Dr. Burke has expertise in the prevention and control of infectious diseases of global concern, including HIV/AIDS, influenza, dengue, and emerging infectious diseases. He is an Institute of Medicine (IOM) member (now the National Academy of Medicine) and has served on previous National Research Council and IOM committees, including the Committee on the Special Immunizations Program for Laboratory Personnel Engaged in Research on Countermeasures for Select Agents and the Committee on Assessment of Future Scientific Needs for Live Variola Virus. Dr. Burke received his B.A. from Western Reserve University and his M.D. from Harvard Medical School.

Philip R. Dormitzer is vice president and chief scientific officer for viral vaccines in the Pfizer Vaccine Research and Development Unit. He is a board certified internal medicine physician. After studying anthropology at Harvard College and carrying out a field study of the Efe Pygmies in the Ituri Forest of Zaire, he completed his M.D. and Ph.D. in Cancer Biology at Stanford University. Dr. Dormitzer completed house-staff training in Internal Medicine at Massachusetts General Hospital and a fellowship in the Harvard Combined Infectious Diseases Training Program. As an assistant professor of pediatrics at Harvard Medical School, Dr. Dormitzer led a structural virology laboratory. The Dormitzer group and its collaborators determined the structures of the rotavirus neutralization antigens by nuclear magnetic resonance spectroscopy, X-ray crystallography, and near atomic resolution electron cryomicroscopy. From 2007-2015 Dr. Dormitzer held a series of positions at Novartis Vaccines and Diagnostics, and was global head of research and vice president at a successor company, Novartis Influenza Vaccines. His teams' research and development programs included vaccines targeting influenza, respiratory syncytial virus, cytomegalovirus, HIV, and parvovirus B19. In 2009, he led the research component of the Novartis response to the H1N1 influenza pandemic, supporting the development and licensure of three pandemic influenza vaccines in the most rapid vaccine response in history. In a Biomedical Advanced Research and Development Authority–funded collaboration with the J. Craig Venter Institute and Synthetic Genomics Vaccines, Inc., the Novartis influenza vaccine research team developed a process to synthesize influenza vaccine seed viruses and deployed the technology in response to the H7N9 influenza outbreak in China. The team's other technology platforms included structurally engineered antigens, adjuvants that target toll-like receptors, and self-replicating messenger RNA vaccines.

Baruch Fischhoff is the Howard Heinz University Professor in the departments of Social and Decision Sciences and of Engineering and Public Policy at Carnegie Mellon University, where he heads the Decision Sciences major. A graduate of the Detroit Public Schools, he holds a B.S. in mathematics and psychology from Wayne State University and an M.A. and a Ph.D. in psychology from the Hebrew University of Jerusalem. He is a member of the National Academy of Medicine and is past President of the Society for Judgement and Decision Making and of the Society for Risk Analysis, and recipient of its Distinguished Achievement Award. He was founding chair of the Food and Drug Administration Risk Communication Advisory Committee and recently chaired the National Research Council Committee on Behavioral and Social Science Research to Improve Intelligence Analysis for National Security. Dr. Fischhoff currently co-chairs the National Research Council Committee on Future Research Goals and Directions for Foundational Science in Cybersecurity and the National Academy of Sciences' Sackler Colloquium on "The Science of Science Communication." He is a former member of the Eugene, Oregon, Commission on the Rights of Women, Department of Homeland Security's Science and Technology Advisory Committee, the World Federation of Scientists Permanent Monitoring Panel on Terrorism, and the Environmental Protection Agency Science Advisory Board, where he chaired the Homeland Security Advisory Committee. He is a fellow of the American Psychological Association, the Association for Psychological Science (previously the American Psychological Society), the Society of Experimental Psychologists, and the Society for Risk Analysis.

Charles N. Haas is the L.D. Betz Chair Professor of Environmental Engineering and Head of the Department of Civil, Architectural, and Environmental Engineering at Drexel University. His broad research interests include drinking water treatment, bioterrorism and risk assessment. Specific research activities include assessment of risks from exposures to deliberately released agents; engineering analysis and optimization of chemical decontamination schemes; microbiological risks associated with pathogens in drinking water, biosolids, and foods; novel kinetic models for disinfection processes and process control; and use of computational fluid dynamics for process modeling. Dr. Haas was co-director of the Center for Advancing Microbial Risk Assessment that is jointly funded by the Department of Homeland Security and the Environmental Protection Agency. He received his Ph.D. from the University of Illinois. He is a past member of the National Academies of Sciences, Engineering, and Medicine's Water Science and Technology Board. He is currently a fellow of multiple societies, including the American Association for the Advancement of Science, American Academy of Microbiology and Society for Risk Analysis.

Michelle M. Mello is Professor of Law at Stanford Law School and Professor of Health Research and Policy at Stanford University School of Medicine. She conducts empirical research into issues at the intersection of law, ethics, and health policy. She is the author of more than 140 articles and book chapters on the medical malpractice system, medical errors and patient safety, public health law, research ethics, the obesity epidemic, pharmaceuticals, and other topics. From 2000-2014, Dr. Mello was a professor at Harvard School of Public Health, where she directed the School's Program in Law and Public Health. In 2013-2014 she completed a Lab Fellowship at Harvard University's Edmond J. Safra Center for Ethics. Dr. Mello teaches courses in torts and public health law. She holds a J.D.

from the Yale Law School, a Ph.D. in Health Policy and Administration from the University of North Carolina at Chapel Hill, and an M.Phil. from Oxford University, where she was a Marshall Scholar. In 2013, she was elected to the National Academy of Medicine, formerly known as the Institute of Medicine.

Sir John Skehel is a graduate of the University College of Wales, Aberystwyth (1962) and gained his Ph.D. from the University of Manchester (1966). He did research at the University of Aberdeen (1965-1968) and was a Helen Hay Whitney Foundation fellow at Duke University and at the Medical Research Council (MRC) National Institute for Medical Research (NIMR) Mill Hill (1968-1971). He was MRC staff scientist at NIMR from 1971 to 2006, Director of the World Health Organization World Influenza Centre from 1975 to 1993, Head of Infections and Immunity from 1985 to 2006 and Director of the NIMR from 1987-2006. He is a visiting scientist in the Division of Virology at The Crick Institute. His research is on the influenza virus hemagglutinin and neuraminidase membrane glycoproteins and the mechanisms of their receptor binding, membrane fusion and enzymic activities. He is a Trustee of the Animal Health Trust. He was elected member of the European Molecular Biology Organization in 1983, fellow of the Royal Society in 1984, member of the Academia Europaea in 1992 and fellow of the Academy of Medical Sciences in 1998 (vice president from 2001-2006) and a foreign associate of the United States National Academy of Sciences in 2014. He was knighted in 1996. He was honorary professor of virology at Glasgow University, Liverpool John Moores University in 2007 and University of Padua (medicine and surgery) in 2010. He is a fellow of the University of Wales and an honorary member of the Society for General Microbiology.

Appendix C

Symposium Agenda

National Academy of Sciences Building
2101 Constitution Avenue NW
Washington, DC 20418
March 10-11, 2016

Thursday, March 10: Overview and Context

8:00 am	**Registration** (coffee and tea will be served)
8:45	**Welcome and Opening Remarks** Moderator: Harvey Fineberg, Symposium Planning Committee Chair *Ralph J. Cicerone,* President, National Academy of Sciences *Margaret Hamburg,* Foreign Secretary, National Academy of Medicine *Jo Handelsman,* Office of Science and Technology Policy *Carrie Wolinetz,* National Institutes of Health
9:15	**Overview of the Draft NSABB Policy Framework and Key Policy Questions** Moderator: Harvey Fineberg, Symposium Planning Committee Chair **Overview of the NSABB Working Paper** *Samuel Stanley,* Stony Brook University and NSABB Chair *Harvey Fineberg,* Symposium Planning Committee Chair **Open Discussion**

10:45 **Break**

11:15 **Informing the Policy Framework: The Risk/Benefit Assessment**
Moderator: Charles Haas, Symposium Planning Committee Member

Lessons from the Risk/Benefit Assessment
Rocco Casagrande, Gryphon Scientific

Comments
Louis (Tony) Cox, Cox Associates
Adam Finkel, University of Pennsylvania
Kara Morgan, Battelle

Open Discussion

12:45 pm **Lunch**
(seating available in the West Court and Members Room - follow signs)

1:45 **The Policy Landscape: United States**
Moderator: Michelle Mello, Symposium Planning Committee Member

Discussants
Gerald Epstein, Department of Homeland Security
Richard Frothingham, Duke University
Lawrence Kerr, Department of Health and Human Services
Philip Potter, St. Jude Children's Research Hospital

Open Discussion

3:15 **Break**

3:45 **The Policy Landscape: International Dimensions of GOF Research**
Moderator: Barry Bloom, Symposium Planning Committee Member

Discussants
Ruxandra Draghia-Akli, European Commission
Keiji Fukuda, World Health Organization

Volker ter Meulen, European Academies Science Advisory Council
Silja Vöneky, University of Freiburg and German Ethics Council

Open Discussion

5:15 **Adjourn**
Reception follows in the Great Hall - all participants welcome

Friday, March 11: Digging Deeper: Key Issues for U.S. Policy Choices

8:30 am **Registration**
(coffee and tea will be served)

9:00 **Informing Policy Design: Insights from the Science of Safety and the Science of Public Consultation**
Moderator: Baruch Fischhoff, Symposium Planning Committee Member

Discussants
Ruthanne Huising, McGill University
Gavin Huntley-Fenner, Huntley-Fenner Advisors
Monica Schoch-Spana, UPMC Center for Health Security

Open Discussion

10:30 **Break**

11:00 **Best Practices to Inform National Policy Design and Implementation: Perspectives of Key Stakeholders in the Biomedical and Public Health Communities**
Moderator: Philip Dormitzer, Symposium Planning Committee Member

Discussants
Michael Callahan, Massachusetts General Hospital and Harvard Medical School
Robert Fisher, U.S. Food and Drug Administration
Jonathan Moreno, University of Pennsylvania
Ethan Settembre, Seqirus

Open Discussion

12:30 pm **Lunch**
(seating available in the West Court and Members Room - follow signs)

1:30 **International Governance: Opportunities for Harmonizing GOF Research Policy and Practice**
Moderator: Ronald Atlas, Symposium Planning Committee Member

Discussants
Nisreen AL-Hmoud, Royal Scientific Society of Jordan
George F. Gao, Chinese Academy of Sciences and China CDC
Gabriel Leung, University of Hong Kong
Michael Selgelid, Monash University
Herawati Sudoyo, Indonesian Academy of Sciences and Eijkman Institute for Molecular Biology[1]

Open Discussion

3:00 **Break**

3:30 **Summing Up**
Moderator: Harvey Fineberg, Symposium Planning Committee Chair

Summary Remarks
Brief remarks from the moderators of the plenary sessions to summarize what emerged from the discussions during the symposium to inform the NSABB's recommendations and the U.S. government's policy choices.

Open Discussion

Concluding Remarks
Harvey Fineberg, Symposium Planning Committee Chair

5:00 **Adjourn**

[1] Dr. Sudoyo was unable to take part in the symposium.

Appendix D

Speaker and Panelist Biographies

Nisreen AL-Hmoud obtained a Ph.D. in microbiology from Abertay University, Dundee, Scotland in 2002. In 2003, she joined the Royal Scientific Society (RSS) of Jordan as a researcher, and since 2009, she has been leading the group of Biosafety at RSS. Dr. AL-Hmoud is a member of the National Biosafety Committee and the National Committee for Science and Technology Ethics in Jordan. She also served as president of the Biosafety and Biosecurity International Consortium (BBIC) steering committee between May 2010 and July 2012. In October 2015, Dr. AL-Hmoud was appointed as director of the Centre for Excellence in Biosafety, Biosecurity and Biotechnology at RSS. Dr. AL-Hmoud started her teaching career in October 2006 as a visiting lecturer of medical microbiology at the department of Biology, Faculty of Science, at the University of Jordan. In February 2008, she joined Princess Sumaya University for Technology (PSUT) as an assistant professor, and later on as a department head and coordinator for the master's program of Environmental Technology and Management. She is also a lecturer at the Health and Community Development Program of Jordan and the School for International Training Study Abroad Program.

Michael Callahan is a physical scientist boarded in both internal medicine and infectious diseases and is a Diplomat of Mass Casualty Care and Tropical Medicine and Hygiene (UK). Dr. Callahan received his M.S. in International Public Health and his M.D. from the University of Alabama School of Medicine, where he was the 19th Tinsley Harrison Scholar and received three academic and research awards in his graduate and medical training. His biodefense clinical research is focused on vaccine defeat, immune evade and multidrug-resistant organisms, and on best practices for highly dangerous pathogen infections in Africa where he prospectively enrolls cutaneous anthrax in Nigeria; and monkey pox, Ebola and Marburg in the Democratic Republic of the Congo and Angola. In 2002, he was appointed clinical director for Cooperative Threat Reduction programs at six former Soviet Union (ex) Biological Weapons Institutes (VECTOR, State Research Center for Applied Microbiology, Kirov, Bersk, RCMDT, Highly Pure, and RIHOP), which included redirecting of unanticipated dual use and gain-of-function programs. From 2005 to 2012, Dr. Callahan led the Defense Advanced Research Projects Agency (DARPA) biodefense therapeutics portfolio, which he expanded from $61 million to $260 million per annum in 2011, involving eight programs that generated nine investigational new drugs (INDs) and three new drug applications with products in market. While at DARPA he launched the Department of Defense Icon program Accelerated Manufacture of Pharmaceuticals (AMP), for which he received the 2010 DARPA Achievement Award, and which generated emergency use good manufacturing practice pH1N1 vaccines, and Nicotinia-expressed monoclonals such as ZMapp. Also while at DARPA, he launched Prophecy, the international physician Early Alert network, which delivers 24/7 emergency consultation, reagents and therapeutics for catastrophic (mass-casualty or HDP) infectious disease outbreaks, severe acute respiratory syndrome Hong Kong and H7N9 Nanjing. His drugs in market include Ambisome (Gilead), which has generated $6 billion since approval, cPG100, and four private-sector INDs involving novel anti-infectives, cytotherapeutics, or host-based antivirals. Dr. Callahan is president of United Therapeutics (UTHR) Division of Cell Therapeutics, and maintains faculty appointments at Massachusetts General Hospital/Harvard Medical School and King Chulalongkom Medical University in Bangkok. Dr. Callahan continues his federal service as infectious disease and biosafety SME to the Academies, the National Security Council, BSEG, the Office of Net Assessment, National Institute of Allergy and Infectious Diseases, MITRE, American Society of Microbiology, Infectious Disease Society of America, and the American Society of Tropical Medicine and Hygiene.

Rocco Casagrande is the Managing Director of Gryphon Scientific, LLC. His projects at Gryphon Scientific focus on bringing rigorous scientific analysis to problems of homeland defense. For the past dozen years, Dr. Casagrande has led more than 50 projects to evaluate and improve U.S. preparedness efforts for a chemical, biological, radiological, and nuclear attack or emerging infectious disease event and to support a better understanding of the threat. Dr. Casagrande also served as the principal investigator of several projects supporting the U.S. government's stance on emerging biotechnologies, including the guidance to the synthetic DNA industry and its moratorium on funding research involving engineered influenza viruses. From December 2002 to March 2003, Dr. Casagrande served as an United Nations Monitoring, Verification and Inspection Commission (UNMOVIC) biological weapons inspector in Iraq where he acted as the chief of the United Nations biological analysis laboratory. Prior to working for UNMOVIC, Dr. Casagrande worked in private industry as an inventor in a nano/biotechnology company. Dr. Casagrande holds a B.A. in chemistry and biology from Cornell University, where he graduated magna cum laude, and a Ph.D. in biology from the Massachusetts Institute of Technology.

Ralph J. Cicerone is the president of the National Academy of Sciences. His research in atmospheric chemistry, climate change, and energy has involved him in shaping science and environmental policy at the highest levels nationally and internationally. Dr. Cicerone was educated at the Massachusetts Institute of Technology (B.S. in electrical engineering) and the University of Illinois at Champaign-Urbana (M.S., Ph.D. in electrical engineering, with a minor in physics). In his early career, he was a research scientist and held faculty positions in electrical and computer engineering at the University of Michigan. The Ralph J. Cicerone Distinguished University Professorship of Atmospheric Science was established there in his honor in 2007. In 1978, he joined the Scripps Institution of Oceanography at the University of California, San Diego, as a research chemist. From 1980 to 1989, he was a senior scientist and director of the Atmospheric Chemistry Division at the National Center for Atmospheric Research in Boulder, Colorado. In 1989, he joined the University of California, Irvine, where he was founding chair of the Department of Earth System Science and was appointed the Daniel G. Aldrich Professor of Earth System Science. As dean of the School of Physical Sciences from 1994 to 1998, he recruited outstanding faculty and strengthened the school's curriculum and outreach programs. Immediately prior to his election as Academy president, Dr. Cicerone served as Chancellor of University of California, Irvine, from 1998 to 2005, a period marked by a rapid rise in the academic capabilities of the campus. His research has focused on atmospheric chemistry, the radiative forcing of climate change due to trace gases, and the sources of atmospheric methane, nitrous oxide, and methyl halide gases.

Louis "Tony" Cox is president of Cox Associates, a Denver-based applied research company specializing in quantitative health risk analysis, casual modeling, advanced analytics, and operations research. Since 1986, Cox

Associates' mathematicians and scientists have applied computer simulation, biomathematical models, biostatistical and epidemiological risk analyses, casual data mining, machine learning, biomathematical modeling and bioinformatics, operations research, and artificial intelligence models to measurably improve health and engineering risk assessment and decision making for public and private sector clients. In 2006, Cox Associates was inducted into the Edelman Academy of the Institute for Operations Research and Management Science (INFORMS), recognizing outstanding real-world achievements in the practice of operations research and the management sciences. In 2012, Dr. Cox was inducted into the National Academy of Engineering (NAE) "for applications of operations research and risk analysis to significant national problems." He is a member of the Academies Board on Mathematical Sciences and their Applications (BMSA) and a member of the Academies Standing Committee on the Use of Public Health Data in Food Safety and Inspection Service Food Safety Programs. Dr. Cox holds a Ph.D. in Risk Analysis (1986) and an S.M. in Operations Research (1985), both from the Massachusetts Institute of Technology; an A.B. from Harvard University (1978); and is a graduate of the Stanford Executive Program (1993). He is honorary full professor of Mathematics at the University of Colorado, Denver, where he has lectured on risk analysis, biomathematics, health risk modeling, computational statistics and causality; is on the Faculties of the Center for Computational Mathematics and the Center for Computational Biology; and is clinical professor of Biostatistics and Informatics at the University of Colorado Health Sciences Center. Dr. Cox is editor-in-chief of *Risk Analysis: An International Journal*, is area editor for Real World Application for the *Journal of Heuristics*, and is on the Editorial Board of the *International Journal of Operations Research and Information Systems*. He is an Edelman Laureate of INFORMS, a member of the American Statistical Association (ASA), and a fellow of the Society for Risk Analysis (SRA).

Ruxandra Draghia-Akli joined DG Research and Innovation of the European Commission as health director in 2009. In her position, Dr. Draghia-Akli is constantly seeking to deepen the reach, the breadth, and the depth of Europe's excellence in health research and innovation (R&I). Before joining the European Commission, Dr. Draghia-Akli served as vice president of Research at VGX Pharmaceuticals (now Inovio) and VGX Animal Health. She received an M.D. from Carol Davilla Medical School and a Ph.D. in human genetics from the Romanian Academy of Medical Sciences. She also completed a doctoral fellowship at the University of Rene Descartes in Paris and post-doctoral training at Baylor College of Medicine and served as faculty at Baylor. In 2012, she became an honorary member of the Romanian Academy of Medical Sciences.

Gerald Epstein is a fellow of the American Physical Society and the American Association for the Advancement of Science. He serves on the editorial board for the journal *Biosecurity and Bioterrorism* and has served on the Biological Threats Panel of the National Academy of Sciences' Committee on International Security and Arms Control and the Biological Sciences Experts Group for the Office of the Director of National Intelligence. He also served on the Committee on Science, Security, and Prosperity, which produced the report *Beyond Fortress America: National Security Controls on Science and Technology in a Globalized World.* He received B.S. degrees in physics and electrical engineering from the Massachusetts Institute of Technology and a Ph.D. in physics from the University of California, Berkeley.

Adam Finkel is currently executive director of the Penn Program on Regulation at the University of Pennsylvania, where he is also a senior fellow at the Penn Law School, and is clinical professor of Environmental Health Sciences at the University of Michigan School of Public Health. From 2004 to 2007, he was a visiting professor of Public and International Affairs at the Woodrow Wilson School at Princeton University. From 2000 to 2003, Dr. Finkel was regional administrator for the U.S. Occupational Safety and Health Administration (OSHA) in Denver, Colorado, responsible for regulatory enforcement, compliance assistance, and outreach activities in the six-state Rocky Mountain region (Region VIII). From 1995 to 2000, he was Director of Health Standards Programs at OSHA headquarters, and was responsible for promulgating and evaluating regulations to protect the nation's workers from chemical, radiological, and biological hazards. Dr. Finkel holds an Sc.D. in environmental health sciences from the Harvard School of Public Health, an M.A. in public policy from Harvard's John F. Kennedy School of Government, an A.B. in biology from Harvard College, and is a certified industrial hygienist. Dr. Finkel has pioneered methodological improvements in human health risk assessment and cost–benefit analysis for the past 25 years, primarily in the areas of quantitative uncertainty analysis, accounting for interindividual variability in susceptibility, and designing regulatory processes to maximize stakeholder input and shed light on economic impacts. He is co-author of four books, including the 2014 volume *Does Regulation Kill Jobs?* In 2006, he received the David P. Rall Award for Advocacy in Public Health from the American Public Health Association for "a career in advancing science in the service of public health protection." In 2013, he received the Alumni Leadership in Public Health Practice Award from the Harvard School of Public Health.

Robert Fisher is director, Regulatory Science for the Food and Drug Administration's (FDA's) Office of Counterterrorism and Emerging

Threats (OCET) and the Medical Countermeasures Initiative (MCMi). He leads the MCMi Regulatory Science Program, oversees intra- and extramural research programs, and works with FDA Centers, PHEMCE stakeholders, and other U.S. and international partners on medical countermeasure-related regulatory science issues. Dr. Fisher joined FDA's Center for Biologics Research and Review (CBER) as a staff fellow in 2006, and served as a staff scientist from 2013-2015. During his tenure at CBER, he provided scientific leadership for regulatory review of chemical, biological, radiological, and nuclear (CBRN) medical countermeasures. He maintained an active research interest in several medical countermeasure related fields, including the modeling of complications related to vaccinia live-virus vaccines and investigating methods for improved characterization of botulism and anthrax antitoxin products. Dr. Fisher received his undergraduate degree in biology from the University of North Carolina at Pembroke and a Ph.D. in toxicology from the University of North Carolina at Chapel Hill. He studied filovirus and poxvirus pathogenesis under a National Research Council Research Associateship at the U.S. Army Medical Research Institute of Infectious Diseases and holds a certificate in Biohazardous Threat Agents and Emerging Infectious Diseases from Georgetown University.

Richard Frothingham is an associate professor of Medicine at Duke University Medical Center. He received his B.S. from the Massachusetts Institute of Technology and his M.D. from Duke. He completed clinical training programs in Medicine, Pediatrics, and Infectious Diseases and maintains board certification in Infectious Diseases. He is also a Certified Biological Safety Professional. Dr. Frothingham directs the National Institute of Allergy and Infectious Diseases Regional Biocontainment Laboratory at Duke University. This laboratory was built to support research to develop drugs, diagnostics, and vaccines for emerging infections and biological threats. The Frothingham lab studies host responses to tuberculosis with the goal of developing better vaccines and treatments. Dr. Frothingham also provides clinical care to persons with HIV infection. Dr. Frothingham serves as co-chair of the Duke Institutional Biosafety Committee (IBC). The Duke IBC has reviewed and managed biological research with the potential for dual use since 2005.

Keiji Fukuda is special representative for antimicrobial resistance for the director-general at the World Health Organization (WHO). He previously served as the assistant director-general for health security, the special adviser on pandemic influenza to the director-general, and director of the Global Influenza Programme. Before joining WHO, Dr. Fukuda served as the chief of the Epidemiology Unit, Influenza Branch, Centers for Disease

Control and Prevention (CDC) in the United States. He has extensive global and national public health experience with health security and emerging infectious diseases, including field investigations and research, capacity building and preparedness, communications, surveillance, and with international governance and frameworks such as the International Health Regulations, the Pandemic Influenza Preparedness Framework and the Codex Alimentarius. He is currently focusing on shaping the global approach to antimicrobial resistance. Dr. Fukuda is a physician and epidemiologist and received his B.A. from Oberlin College, his M.D. from the University of Vermont, his M.P.H. from the University of California, Berkeley, and additional training in epidemiology at CDC.

George Gao obtained his Ph.D. (D.Phil.) degree in 1995 from Oxford University, United Kingdom. He was selected by the Chinese Academy of Sciences "Hundred Talents" program in 2004, and received the National Natural Science Foundation of China (NSFC) Distinguished Young Scholar title in 2005. He is the chief scientist of two consecutive projects on the mechanism of interspecies transmission of viral pathogens and a leading principal investigator of the NSFC Innovative Research Group. He is also a member of the steering committee for the International Consortium of Anti-Virals (ICAV), and a visiting professor at Oxford University. He was awarded the World Academy of Sciences (TWAS) prize in Medical Sciences in 2012 and the Nikkei Asia Prize in 2014. Dr. Gao is a member of the Chinese Academy of Sciences, a fellow of TWAS, a fellow of the American Academy of Microbiology, and the director and professor in the Chinese Academy of Sciences Key Laboratory of Pathogenic Microbiology and Immunology, Institute of Microbiology. He is also the vice president of the Beijing Institutes of Life Science, Chinese Academy of Sciences, Deputy Director-General of the Chinese Center for Disease Control and Prevention, and dean of the UCAS Cunji College of Medicine.

Margaret Hamburg earned her B.A. from Harvard College, her M.D. from Harvard Medical School, and completed her residency at what is now New York Presbyterian Hospital-Weill Cornell Medical Center. She conducted neuroscience research at Rockefeller University in New York and at the National Institute of Mental Health, and later focused on HIV/AIDS research and policy as assistant director of the National Institute of Allergy and Infectious Diseases. In 1991, after just 1 year in the New York City Department of Health, Dr. Hamburg was named its commissioner. During her 6-year tenure, she implemented rigorous public health initiatives that tackled the city's most pressing crises head-on—including improved services for women and children, a needle-exchange program to combat HIV transmission, and the nation's first public health bioterrorism defense program. The most celebrated achievement during her leadership was her aggressive approach to the city's tuberculosis epidemic, which led to an 86 percent decline in drug-resistant tuberculosis in just 5 years. In 1997, 3 years after she was elected one of the youngest-ever members of the Institute of Medicine, President Bill Clinton named Dr. Hamburg assistant secretary for planning and evaluation in the Department of Health and Human Services, where she served until the end of the Clinton Administration. She then became founding vice president for biological programs at the Nuclear Threat Initiative, a foundation dedicated to reducing the threat to public safety from nuclear, chemical, and biological weapons. President Barack Obama nominated Dr. Hamburg for the post of the Food and Drug Administration commissioner on March 14, 2009. Dr. Hamburg is a member of the National Academy of Medicine and currently serves as its foreign secretary.

Jo Handelsman is the associate director for science at the White House Office of Science and Technology Policy (OSTP), appointed by President Obama and confirmed by the Senate in June of 2014. Dr. Handelsman helps to advise President Obama on the implications of science for the nation, ways in which science can inform U.S. policy, and on federal efforts in support of scientific research. Prior to joining OSTP, Dr. Handelsman was the Howard Hughes Medical Institute Professor and Frederick Phineas Rose Professor in the Department of Molecular, Cellular and Developmental Biology at Yale University. She previously served on the University of Wisconsin, Madison, faculty as a professor in plant pathology from 1985 to 2009 and as professor and chair of the Department of Bacteriology from 2007 to 2009. In 2013, she served as president of the American Society for Microbiology. Dr. Handelsman is an expert in communication among bacteria that associate with soil, plants, and insects and helped pioneer the field of metagenomics, bridging agricultural and medical sciences. Dr. Handelsman is also recognized for her research on science education and women and minorities in science, and received the Presidential Award for Excellence in Science Mentoring in 2011. Dr. Handelsman also co-chaired the PCAST working group that developed the 2012 report, "Engage to Excel," which contained recommendations to the president to strengthen STEM (science, technology, engineering, and mathematics) education to meet the workforce needs of the next decade in the United States. Dr. Handelsman received a B.S. from Cornell University and a Ph.D. in Molecular Biology from the University of Wisconsin, Madison.

Ruthanne Huising is an ethnographer of work and organizations. She studies how organizations respond to external pressures to change and the implications of these changes for professional control and expertise.

Across various projects she has observed how organizations accommodate regulatory change (Canada's Human Pathogens and Toxins Act), auditing fads (Environment, Health & Safety Management Systems), and efficiency efforts (Ontario's perioperative coaching program), and the complex responses of scientists, biosafety officers, health physicists, surgeons, nurses, and administrators. Ruthanne is an associate professor in the faculty of management at McGill University. She received her Ph.D. from the Sloan School of Management at the Massachusetts Institute of Technology.

Gavin Huntley-Fenner is an independent human factors consultant. His consulting and research interests are focused on the contribution of risk perception and reasoning to warnings effectiveness. Prior to focusing full-time as a human factors consultant, Dr. Huntley-Fenner was a business consultant at McKinsey & Company. He began his professional career as an assistant professor at the University of California, Irvine, after earning his Ph.D. in Brain and Cognitive Sciences from the Massachusetts Institute of Technology. From 2010-2014, he served as a member of the Food and Drug Administration's Risk Communication Advisory Committee.

Lawrence Kerr is the director of pandemics and emerging threats within the Office of Global Affairs at the Department of Health and Human Services (HHS). Dr. Kerr leads and manages the office, overseeing a broad policy portfolio including the global health security agenda implementation, pandemics and emerging threats, antimicrobial resistance, security policy issues (biosafety and biosecurity, biothreat prevention [Biological Weapons and Toxins Convention, United Nations Security Council 1540, Global Partnership against the Spread of Materials and Weapons of Mass Destruction]), and dual use research of concern. Prior to joining HHS in December 2015, Dr. Kerr served as the director for medical preparedness policy in the Resilience Directorate at the White House National Security Council Staff as the principal staff member responsible for coordinating policy regarding public health and medical resilience for biological events, whether the results of naturally emerging disease or deliberate release including his role on the Ebola Task Force. He previously served as the senior bio advisor to the director of the National Counterproliferation Center (NCPC) within the Office of the Director of National Intelligence. Dr. Kerr advised the senior leadership on strategic plans to prevent and counter the spread of biological weapons of mass destruction. Before joining NCPC in April 2006, he was director for biodefense policy with the White House Homeland Security Council in the Executive Office of the President. He served as assistant director for homeland security for the Office of Science and Technology Policy (OSTP) and as director of bioterrorism, research and development for the Office of Homeland Security in the Executive Office of the President. Dr. Kerr joined the Life Sciences division of OSTP in January 2001, where he came from his position at the National Institute of Allergy and Infectious Diseases at the National Institutes of Health. He holds a B.S. in Biology and Art History and a Ph.D. in Cell Biology, both from Vanderbilt University.

Gabriel Leung is dean of medicine and chair professor of public health medicine at The University of Hong Kong. Previously he was Hong Kong's first under secretary for food and health, then director of the chief executive's office in government. Dr. Leung is one of Asia's leading epidemiologists, having authored more than 400 scholarly papers and edited numerous leading journals. He directs the university's World Health Organization Collaborating Centre for Infectious Disease Epidemiology and Control. His research defined the epidemiology of two novel viral epidemics, namely severe acute respiratory syndrome coronavirus in 2003 and influenza A (H7N9) in 2013. While in government, he led Hong Kong's policy response against the 2009 influenza A (H1N1) pandemic.

Jonathan Moreno is 1 of 16 Penn Integrates Knowledge university professors at the University of Pennsylvania, holding the David and Lyn Silfen chair. He is also professor of medical ethics and health policy, of history and sociology of science, and of philosophy. Dr. Moreno is a senior fellow at the Center for American Progress in Washington, DC. In 2008-2009 he served as a member of President Obama's transition team. His work has been cited by Al Gore and was used in the development of the screenplay for *The Bourne Legacy*. His online neuroethics course drew more than 36,000 registrants in 2013. Dr. Moreno's writings have been translated into Chinese, German, Japanese, and Portugese. The *American Journal of Bioethics* has called him "the quietly most interesting bioethicist of our time." Dr. Moreno is an elected member of the National Academy of Medicine. He has served as a senior staff member for three presidential advisory commissions, including the current bioethics commission under President Obama, and has given invited testimony for both houses of Congress. Dr. Moreno is the U.S. member of the United Nations Educational, Scientific and Cultural Organization International Bioethics committee. Dr. Moreno received his Ph.D. in philosophy from Washington University in St. Louis, was an Andrew W. Mellon postdoctoral fellow, holds an honorary doctorate from Hofstra University, and is a recipient of the Benjamin Rush Medal from the College of William and Mary Law School and the Dr. Jean Mayer Award for Global Citizenship from Tufts University.

Kara Morgan has 16 years of experience in risk analysis and decision analysis. She earned her B.S. in Mathematics from Michigan State University, her M.S. in Environmental Science from Indiana University, and her Ph.D. in Engineering and Public Policy from Carnegie Mellon University. After earning her Ph.D., she worked for 4 years at Research Triangle Institute, supporting the Environmental Protection Agency with the use of data-based decision making methods. Then, she spent 10 years at the Food and Drug Administration, working to support the development and implementation of risk-based decision-making tools and to implement strategic program planning for improving the achievement of outcomes. She is currently a research leader at Battelle Memorial Institute in the Health and Analytics sector. In that role, she works with clients to improve their use of data to inform decision making, supports knowledge management tasks related to quality measures for health care improvement, and works with clients to assess the outcomes their programs are achieving. She is also an adjunct professor at the Ohio State University's Glenn College of Public Affairs, where she teaches courses on risk and decision analysis. Dr. Morgan's professional focus has been on developing tools and methods for supporting effective data-driven risk management decisions. Her areas of emphasis include performance measurement, strategic planning, program evaluations, knowledge management, risk and decision analysis, and application of these tools and methods to improve decision making and improve outcomes.

Philip Potter obtained his Ph.D. in molecular carcinogenesis at the Paterson Institute for Cancer Research in Manchester, United Kingdom, and moved to St. Jude Children's Research Hospital in Memphis, Tennessee, shortly thereafter. His laboratory has worked for many years on the modulation of the response of tumor cells to chemotherapy, using both small molecule and molecular approaches. The latter has principally involved the use of adenovirus to deliver agents, such as ribozymes and drug metabolizing enzymes to cells, both in vitro and in vivo. Consequently, he has expertise in the design and construction of viral vectors and their practical use in the laboratory. Dr. Potter has more than 11 years of experience serving on the St. Jude Institutional Biosafety Committee, including as vice chairman and chairman. He is currently the vice chair of the IBC, and the chairman of the dual use research of concern subcommittee for the institution.

Monica Schoch-Spana, a medical anthropologist, is a senior associate with the University of Pittsburgh Medical Center (UPMC) Center for Health Security and a faculty member with the School of Medicine at the University of Pittsburgh and the Department of Anthropology at Texas State University. Dr. Schoch-Spana is a leading social science researcher in public health emergency preparedness. Her studies have been influential in debunking myths about mass behaviors in the context of bioterrorism and other health crises and in reframing the management of catastrophic health events to include social, ethical-moral, and governance dimensions. National advisory roles include serving on the Homeland Security Subcommittee of the Board of Scientific Counselors for the Environmental Protection Agency, the Resilient America Roundtable of the National Academy of Sciences, and the National Research Council Committee on Increasing National Resilience to Hazards and Disasters. Dr. Schoch-Spana has chaired national working groups to produce peer-reviewed, evidence-based consensus guidance for authorities on how to partner with citizens and civil society in relation to bioterrorism response, influenza pandemic planning, and nuclear incident preparedness, and she has organized three national meetings on how to strengthen community resilience to extreme health events. Her current research projects focus on local health department capacity for community engagement, communication dilemmas concerning medical countermeasures, and public participation in the development of policies for allocating scarce medical resources in a disaster. In 2003, Dr. Schoch-Spana helped establish the UPMC Center for Health Security. Prior to that, she worked at the Johns Hopkins University Center for Civilian Biodefense Strategies starting in 1998. She received her Ph.D. in cultural anthropology from Johns Hopkins University and a B.A. from Bryn Mawr College.

Ethan Settembre is vice president and head of research for Seqirus. He holds a Ph.D. in biochemistry from Cornell University and completed his postdoctoral training in Structural Virology at Harvard Medical School. He then joined Novartis Vaccines & Diagnostics in 2008 where he held several key positions in research developing vaccines against multiple viral targets, including influenza. Currently, he heads the Seqirus Research group focused on influenza vaccine development.

Michael Selgelid is director of the Centre for Human Bioethics, and the World Health Organization (WHO) Collaborating Centre for Bioethics therein, at Monash University in Melbourne, Australia. He is a member of the board of directors of the International Association of Bioethics and serves on the Ethics Review Board of Médecins Sans Frontières. His main research focus is public health ethics with emphasis on ethical issues associated with infectious disease. He edits a book series in Public Health Ethics Analysis for Springer and a book series in Practical Ethics and Public Policy for ANU Press. He is co-editor of *Monash Bioethics Review* and an associate editor of the *Journal of Medical Ethics*. Dr. Selgelid earned a B.S. in

Biomedical Engineering from Duke University, and a Ph.D. in Philosophy from University of California, San Diego.

Samuel L. Stanley, Jr. was appointed as the fifth president of Stony Brook University in May 2009. Since that time he has presided over a tremendous growth of the university, through the implementation of a faculty hiring program that has brought 200 net new faculty to Stony Brook, a 5-fold increase in endowed professorships, the largest number of applicants and most accomplished classes in the school's history, and record fundraising totals, including one of the largest gifts ever to a public university. Before becoming president of Stony Brook University, Dr. Stanley served as vice chancellor for research at Washington University in St. Louis, where he had a distinguished career as a biomedical researcher with a focus on host defense against emerging pathogens. Dr. Stanley currently serves as the chair of the National Science Advisory Board for Biosecurity (NSABB), is a member of the National Security Higher Education Advisory Board (NSHEAB), is the chair of Brookhaven Science Associates (BSA), which manages Brookhaven National Laboratory, is a member the board of directors of Cold Spring Harbor Laboratory, and is a member of the board of directors of the Research Foundation, State University of New York.

Volker ter Meulen qualified as an M.D. in 1960. He received his post-doctoral training in virology in the United States, at the Children's Hospital of Philadelphia. On returning to Germany in 1996, he specialized in pediatrics and was subsequently visiting scientist at the Wistar Institute for Anatomy and Biology in Philadelphia and at the Viral and Rickettsial Disease Laboratory in Berkeley, from 1969-1970. In 1975 he became a full professor and chairman of the Institute of Virology and Immunobiology at the University of Würzburg. He retired in 2002, having twice been elected dean of the faculty of medicine at Würzburg University. During his research career, Dr. ter Meulen worked on molecular and pathogenic aspects of viral infections in man and animals, in particular on infections of the central nervous system. Internationally, Dr. ter Meulen has served on a number of committees of organisations and scientific societies/unions in the area of virology and infectious diseases, covering a broad spectrum of important issues connected to human and animal pathogens. From 2003-2010, Dr. ter Meulen was president of the German Academy of Sciences Leopoldina. From 2007-2010, he was president of the European Academies Science Advisory Council (EASAC), the association of the National Science Academies of the European Union, which is the IAP associated regional network for Europe. He was elected IAP Co-Chair in February 2013.

Silja Vöneky is co-director of the Institute for Public Law and is a professor of public international law, comparative law, and ethics of law at the University of Freiburg (Germany). Her areas of focus include international law, international humanitarian law, international environmental law, the law of the sea, international protection of human rights, the relation of ethics and law, and especially questions on how to regulate existential risks (biosecurity law and democratic legitimacy.) Since 2001, Professor Vöneky has served as the legal adviser to the German Federal Foreign Office, German Federal Ministry of Research, German Federal Ministry of the Environment, and the Alfred Wegener Institute for Scientific Marine and Polar Research. Since 2012, she has been a member of the German Ethics Council, appointed on the proposal of the federal government, and was the head of the Working Group on Biosecurity of the German Ethics Council.

Carrie Wolinetz is associate director for science policy and director of the Office of Science Policy (OSP) at the National Institutes of Health (NIH). As leader of OSP, she advises the NIH director on science policy matters of significance to the agency, the research community, and the public on a wide range of issues, including human subjects protections, biosecurity, biosafety, genomic data sharing, regenerative medicine, the organization and management of NIH, and the outputs and values of NIH-funded research. Prior to joining NIH, Dr. Wolinetz worked on biomedical research policy issues as the deputy director for federal affairs at the Association of American Universities (AAU) and the director of scientific affairs and public relations at the Federation of American Societies for Experimental Biology (FASEB). She also served as the president of United for Medical Research, a leading NIH advocacy coalition. Outside of NIH, Dr. Wolinetz teaches as an adjunct assistant professor at Georgetown University in the School of Foreign Service's program on Science, Technology & International Affairs. She has a B.S. in animal science from Cornell University, and she received her Ph.D. in animal science from Pennsylvania State University, where her area of research was reproductive physiology.

Appendix E

List of Attendees

Nisreen AL-Hmoud
Royal Scientific Society of Jordan

Abdulaziz Alagaili
King Saud University

Lida Anestidou
National Academies of Sciences, Engineering, and Medicine

Ronald Atlas
University of L ouisville

Rachel Bartholomew
Pacific Northwest National Laboratory

Kavita Berger
Gryphon Scientific

Kenneth Berns
University of Florida

Lizbet Boroughs
Association of American Universities

Donald Burke
University of Pittsburgh

Michael Callahan
Massachusetts General Hospital
Harvard Medical School

Elizabeth Cantwell
Lawrence Livermore National Laboratory

Sarah Carter
J. Craig Venter Institute

Rocco Casagrande
Gryphon Scientific

Ralph Cicerone
National Academy of Sciences

Louis "Tony" Cox
Cox Associates

Bruce Crise
Institutional Biosafety Committee Consultant

Genevieve Croft
Association of Public and Land-grant Universities

Patricia Delarosa
National Institute of Allergy and Infectious Diseases

Diane DiEuliis
National Defense University

Philip Dormitzer
Pfizer Vaccine Research and Development

Ruxandra Draghia-Akli
European Commission

David Drew
Woodrow Wilson Center

Leo Einck
EpiVax, Inc.

Gerald Epstein
Department of Homeland Security

Nicholas Evans
University of Pennsylvania

Harvey Fineberg
Gordon and Betty Moore Foundation

Adam Finkel
University of Pennsylvania

Andi and Baruch Fischhoff
Carnegie Mellon University

Robert Fisher
Food and Drug Administration

J. Patrick Fitch
National Biodefense Analysis and Countermeasures Center

Meg Flanagan
Department of State

Greg Frank
Infectious Diseases Society of America

Richard Frothingham
Duke University

Keiji Fukuda
World Health Organization

George F. Gao
Chinese Academy of Sciences
Chinese Center for Disease Control and Prevention

Liz Geltman
CUNY School of Public Health

Daniel Feakes
Biological Weapons Convention

Ashley Grant
Government Accountability Office

Gigi Gronvall
University of Pittsburgh Medical Center Center for Health Security

Charles Haas
Drexel University

Margaret Hamburg
National Academy of Medicine

Marie-Louise Hammarskjöld
University of Virginia

Jo Handelsman
Office of Science and Technology Policy
The White House

Chris Hanson
National Institute of Allergy and Infectious Diseases

Brit Hart
Association of Public Health Laboratories

Jack Herrmann
National Academies of Sciences, Engineering, and Medicine

Rona Hirschberg
Rona Hirschberg Consulting LLC

Andrea Hodges
National Academies of Sciences, Engineering, and Medicine

Ruthanne Huising
McGill University

Gavin Huntley-Fenner
Huntley-Fenner Advisors

Jo L. Husbands
National Academies of Sciences, Engineering, and Medicine

Tom Inglesby
University of Pittsburgh Medical Center Center for Health Security

Katherine Jones
Sandia National Laboratories

Joseph Kanabrocki
University of Chicago

Luba Katz
Abt Association

Lawrence Kerr
Department of Health and Human Services

Andy Kilianski
Edgewood Chemical Biological Center

Tamika Knight
Department of Homeland Security

Gregory Koblentz
George Mason University

Margaret Kosal
Georgia Tech

Amy Krafft
National Institute of Allergy and Infectious Diseases
National Institutes of Health

Viktoriya Krakovna
Future of Life Institute

Jan Leach
Colorado State University

James LeDuc
University of Texas Medical Branch

Betty Lee
Department of Commerce

Gabriel Leung
The University of Hong Kong

Rachel Levinson
Arizona State University

Carol Linden
Office of the Chief Scientist
Food and Drug Administration

Marc Lipsitch
Harvard T.H. Chan School of Public Health

Daniel Lucey
Georgetown University

Francis Macrina
Virginia Commonwealth University

Joanne Manrique
Center for Global Health and Diplomacy

Monique Mansoura
Seqirus

Brendan McGovern
National Academies of Sciences, Engineering, and Medicine

Michelle Mello
Stanford University School of Medicine

Carissa Meyer
Gryphon Scientific

Robin Miller
National Academies of Sciences, Engineering, and Medicine

Piers Millett
Biosecure Ltd.

Allison Mistry
Gryphon Scientific

B. J. Mitchell
Renaissance Today

Jonathan Moreno
University of Pennsylvania

Kara Morgan
Battelle

Rebecca Moritz
University of Wisconsin, Madison

Stephen Morse
Columbia University

Anna Muldoon
Office of the Assistant Secretary of Preparedness and Response
Department of Health and Human Services

Stuart Nightingale
Office of Science Policy
National Institutes of Health

Jenna Ogilvie
National Academies of Sciences, Engineering, and Medicine

Marina O'Reilly
Office of Science Policy
National Institutes of Health

Megan Palmer
Stanford University

Atsuko Polzin
Seqirus

Philip Potter
St. Jude Children's Research Hospital

Philip Price
Wellcome Trust

Kevin Ramkissoon
National Institutes of Health

Catherine Rhodes
University of Cambridge

Ryan Ritterson
Gryphon Scientific

Ben Rusek
National Academies of Sciences, Engineering, and Medicine

Frederik Schagen
Netherlands Commission on Genetic Modification

Monica Schoch-Spana
University of Pittsburgh Medical Center Center for Health Security

Michael Selgelid
Monash University

Sarath Seneviratne
Food and Drug Administration

Aanika Senn
National Academies of Sciences, Engineering, and Medicine

Ethan Settembre
Seqirus

Frances Sharples
National Academies of Sciences, Engineering, and Medicine

Michael Shaw
Centers for Disease Control and Prevention

Ian Simon
Institute for Defense Analyses Science and Technology Policy Institute

Katherine Sixt
Institute for Defense Analyses

Robert Sorenson
Department of State

Erin Sorell
George Washington University

David Spiro
National Institutes of Health

Samuel Stanley, Jr.
Stony Brook University

John Steel
Emory University

Erik Stemmy
National Institute of Allergy and Infectious Diseases

Kata Subbarao
National Institute of Allergy and Infectious Diseases

Volker ter Meulen
European Academies Science Advisory Council

Krista Versteeg
National Academies of Sciences, Engineering, and Medicine

Christopher Viggiani
National Institutes of Health

Silja Vöneky
University of Freiburg/German Ethics Council

James Welch
Elizabeth R. Griffin Foundation

Susan Wolf
University of Minnesota

Carrie Wolinetz
National Institutes of Health

Scott Wollek
National Academies of Sciences, Engineering, and Medicine

David Woodland
Keystone Symposia